Armand Marie Leroi is Reader in Evolutionary
Developmental Biology at Imperial College London.
Mutants was shortlisted for the Aventis Prize and
won the Guardian First Book Award.

For automatic updates on Armand Marie Leroi visit
harperperennial.co.uk and register for AuthorTracker.

From the reviews of *Mutants*:

'One of the many strengths of this complex but accessible study is
its combination of medical history and an admirably clear
exposition of up-to-date scientific thinking. Essentially, this book
tells the story of the development of an individual from embryo to
old age. It does so by exploring what genetic mutations reveal
about the growth of different parts of the body, and by combining
fascinating narratives with sophisticated science ... Poetic,
philosophical, profound, witty and challenging'   *Guardian*

MUTANTS

On the Form, Varieties and Errors
of the Human Body

ARMAND MARIE LEROI

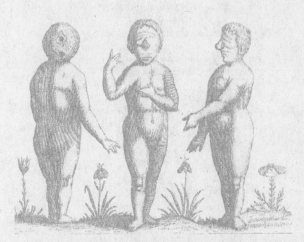

HARPER PERENNIAL
London, New York, Toronto and Sydney

Harper Perennial
An imprint of HarperCollins*Publishers*
77–85 Fulham Palace Road
Hammersmith
London W6 8JB

www.harperperennial.co.uk

This edition published by Harper Perennial 2005
1

First published by HarperCollins*Publishers* 2003

A catalogue record for this book
is available from the British Library

ISBN 978 0 00 653164 7

Set in Granjon

CONTENTS

ILLUSTRATIONS

I

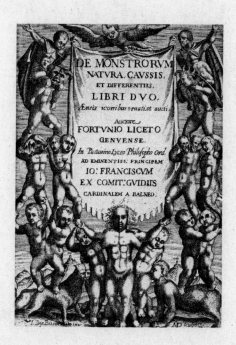

MUTANTS

[An introduction]

We had heard that a monster had been born at Ravenna, of which a drawing was sent here; it had a horn on its head, straight up like a sword, and instead of arms it had two wings like a bat's, and the height of its breasts it had a fio [Y-shaped mark] on one side and a cross on the other, and lower down at the waist, two serpents, and it was a hermaphrodite, and on the right knee it had an eye, and its left foot was like an eagle. I saw it painted, and anyone who wished could see this painting in Florence.

It was March 1512, and a Florentine apothecary named Lucca Landucci was writing up his diary. He had much to write about. Northern Italy was engulfed by war. Maximillian of Germany and Louis XII of France were locked in combat with the Spanish, English and Pope Julius II for control of the Venetian Republic. City after city was ravaged as the armies

FRONTISPIECE TO FORTUNIO LICETI 1634 *De monstrorum natura caussis et differentiis.*

3

traversed the campagna. Ravenna fell eighteen days after the monster's birth. 'It was evident,' wrote Landucci, 'what evil the monster had meant for them! It seems as if some great misfortune always befalls the city when such things are born.'

THE MONSTER OF RAVENNA (1512). FROM ULISSE ALDROVANDI 1642 *Monstrorum historia*.

Landucci had not actually seen the monster. It had been starved to death by order of Julius II, and Landucci's account is of a drawing that was on public display in Florence. That image was among the first of many. Printed woodcuts and engravings spread the news of the monster throughout Europe, and as they spread, the monster acquired a new, posthumous, existence. When it left Ravenna it had two legs; by the time it arrived in Paris it had only one. In some prints it had bat wings, in others they were more like a bird's; it had hermaphrodite genitalia or else a single large erection. It became mixed up with the images of another monster born in Florence in 1506, and then fused with a medieval icon of sinful humanity called 'Frau Welt' – a kind of bat-winged, single-legged Harpy who grasped the globe in her talons.

As the monster travelled and mutated, it also accreted ever more complex layers of meaning. Italians took it as a warning of the horrors of war. The French, making more analytical effort, interpreted its horn as pride, its wings as mental frivolity and inconstancy, its lack of arms as the absence of good works, its raptor's foot as rapacity, and its deformed genitalia as sodomy – the usual Italian vices in other words. Some said that it was the child of a respectable married woman; others that it was the product of a union between a nun and a friar. All this allegorical

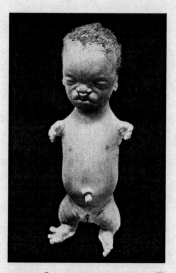

ROBERTS'S SYNDROME. STILLBORN INFANT. FROM B.C. HIRST AND G.A. PIERSOL 1893 *Human monstrosities.*

freight makes it hard to know what the monster really was. But it seems likely that it was simply a child who was born with a severe, rare, but quite unmysterious genetic disorder. One can even hazard a guess at Roberts's syndrome, a deformity found in children who are born with an especially destructive mutation. That, at least, would account for the limb and genital anomalies, if not the two serpents on its waist and the supernumerary eye on its knee.

In the sixteenth and seventeenth centuries, monsters were everywhere. Princes collected them; naturalists catalogued them; theologians turned them into religious propaganda. Scholars charted their occurrence and their significance in exquisitely illustrated books. In Germany, Conrad Lycosthenes produced his *Prodigiorum ac ostentorum chronicon* (1557, later translated as *The Doome, calling all men to judgement*); from France came Pierre Boaistuau's *Histoires prodigieuses* (History of prodigies, 1560–82) and Ambroise Paré's *Des monstres et prodiges* (Monsters and prodigies, 1573). A little later, the Italians weigh in with Fortunio Liceti's *De monstrorum natura caussis et differentiis* (On the nature, causes and differences of monsters, 1616) and Ulisse Aldrovandi's *Monstrorum historia* (History of monsters, 1642).

In an age in which religious feelings ran high, deformity was often taken as a mark of divine displeasure, or at least of a singularly bad time in the offing. Boaistuau's *Histoires prodigieuses*, which is especially rich in demonic creatures, has a fine account not only of the unfortunate Monster of Ravenna but also of the Monster of Cracow – an inexplicably deformed child who apparently entered the world in 1540 with barking dogs' heads mounted on its elbows, chest and knees and departed it four hours later declaiming 'Watch, the Lord Cometh.' Allegory was a sport at which Protestant scholars excelled. In 1523 Martin Luther and Philipp Melanchthon published a pamphlet in which they described a deformed 'Monk-Calf' born in Freiburg and another creature, possibly human, that had been fished out of the Tiber, and interpreted both, in vitriolic terms, as symbols of the Roman Church's corruption. Catholics responded by identifying the calf as Luther.

By the late 1500s, a more scientific spirit sets in. In *Des monstres,* his engagingly eclectic compendium of nature's marvels, the Parisian surgeon Ambroise Paré lists the possible causes of monsters. The first entry is 'The Wrath of God', but God's wrath now seems largely confined to people who have sex with animals (and so produce human-horse/goat/dog/sheep hybrids) or during menstruation (Leviticus disapproved). Luther's Monk-Calf also appears in *Des monstres*, but shorn of its anti-papal trappings. It is, instead, a monster of the 'imagination', that is, one caused by maternal impressions – the notion, prevalent in Paré's day and still in the late nineteenth century, that a pregnant woman can, by looking at an unsightly thing, cause deformity in her child. Like most of the other causes of deformity that Paré proposes (too much or too little semen, narrow wombs, indecent posture), the theory of maternal impressions is simply wrong. But it is rational insofar that it does not appeal to supernatural agents, and *Des monstres* marks the presence of a new idea: that the causes of deformity must be sought in nature.

At the beginning of the seventeenth century, teratology – literally, the 'science of monsters' – begins to leave the world of the medieval wonder-books behind. When Aldrovandi's *Monstrorum historia* was published posthumously in 1642, its mixture of the plausible (hairy people, giants, dwarfs and conjoined twins) and the fantastic (stories taken from Pliny of Cyclopes, Satyrs and Sciapodes) was already old-fashioned. Fortunio Liceti's treatise, published in 1616, is mostly about children with clearly recognisable abnormalities – as can be seen from the frontispiece where they are assembled in heraldic poses. True, they include a calf born with a man's head and, inevitably, the Monster of Ravenna. But even this most terrible of creatures is almost seraphic as it grasps the title-banner in its talons.

There is a moment in time, a few decades around the civil war that racked seventeenth-century England, when the discovery of the natural world has a freshness and clarity that it seems to have lost since. When vigorous prose could sweep away the intellectual wreckage of antiquity and simple experiments could reveal beautiful new truths about nature. In Norfolk, the physician and polymath Sir Thomas Browne published his *Pseudodoxia epidemica, or, enquiries into very many received tenents and commonly presumed truths* (1646). In this strange and recondite book he investigated a host of popular superstitions: that the feathers of a dead kingfisher always indicate which way the wind is blowing, that the legs of badgers are shorter on one side than the other, that blacks were black because they were cursed, that there truly were no rainbows before the Flood – and concludes, in each case, that it isn't so. In another work, his *Religio medici* of 1642, he touches on monsters. There is, he writes, 'no deformity but in Monstrosity; wherein notwithstanding, there is a kind of Beauty. Nature so ingeniously contriving the irregular parts, as they become sometimes more remarkable than the principal Fabrick.' This is not precisely a statement of scientific naturalism, for Browne sees the works of nature – all of them, even the most deformed – as the works of God, and if they are the work of God then they cannot be repugnant. It is, in a few beautiful periods, a statement of tolerance in an intolerant age.

At Oxford, William Harvey, having triumphantly demonstrated the circulation of the blood, was attempting to solve the problem of the generation of animals. In 1642, having declared for the King, Harvey retreated from the turmoil of civil war by studying the progress of chick embryos using the eggs of a hen that lived in Trinity College. The Italians Aldrovandi and Fabricius had already carried out similar studies, the former being the first to do so since Aristotle. But Harvey had greater ambitions. Charles I delighted in hunting the red deer that roamed, and still roam, the Royal Parks of England, and he allowed Harvey to dissect his victims. Harvey followed the progress of the deer embryo month by month, and left one of the loveliest descriptions of a mammalian foetus ever written. 'I saw long since a foetus,' he writes, 'the magnitude of a peascod cut out of the uterus of a doe, which was complete in all its members & I showed this pretty spectacle to our late King and Queen. It did swim, trim and perfect, in such a kind of white, most transparent and crystalline moysture (as if it had been treasured up in some most clear glassie receptacle) about the bignesse of a pigeon's egge, and was invested with its proper coat.' The King apparently followed Harvey's investigations with great interest, and it is a poignant thought that when Charles I was executed, England lost a monarch with a taste for experimental embryology, a thing not likely to occur again soon.

The frontispiece of Harvey's embryological treatise, *De generatione animalium* (1651), shows mighty Zeus seated upon an eagle, holding an egg in his hand from which all life emerges. The egg bears the slogan *Ex Ovo Omnia* – from the egg, all – and it is for this claim, that the generation of mammals and chickens and everything else is fundamentally alike, that the work is today mostly remembered, even though Harvey neither used the slogan himself nor proved its truth. Harvey has some things to say about monstrous births. He revives, and queries, Aristotle's claim that monstrous chickens are produced from eggs with two yolks. This may not seem to amount to much, but it was the expression of an idea, dormant for two millennia, that the causes of monstrosity are not just a matter for idle speculation of the sort that Paré and Liceti dealt in, but are instead an experimentally tractable problem.

It was, however, a contemporary of Harvey's who stated the true use of deformity to science – and did so with unflinching clarity. This was Francis Bacon. Sometime Lord Chancellor of England, Bacon comes down to us with a reputation as the chilliest of intellectuals. His ambition was to establish the principles by which the scientific inquiry of the natural world was to be conducted. In his *Novum organum* of 1620 Bacon begins by classifying natural history. There are, he says, three types of natural history: that which 'deals either with the *Freedom* of nature or with the *Errors* of nature or with the *Bonds* of nature; so that a good division we might make would be a history of *Births*, a history of *Prodigious Births*, and a history of *Arts*; the last of which we have also often called the *Mechanical* and the *Experimental* Art'. In other words, natural history can be divided into the study of normal nature, aberrant nature and nature manipulated by man. He then goes on to tell us how to proceed with the second part of this programme. 'We must make a collection or particular natural history of all the monsters and prodigious products of nature, of every novelty, rarity or abnormality.' Of course, Bacon is interested in collecting aberrant objects not for their own sake, but in order to understand the causes of their peculiarities. He does not say *how* to get at the causes – he simply trusts that science will one day provide the means.

Bacon's recommendation that 'monsters and prodigious products' should be collected would not have startled any of his contemporaries. Princes such as Rudolf II and Frederick II of Austria had been assembling collections of marvels since the mid-1500s. Naturalists were at it too: Ulisse Aldrovandi had assembled no fewer than eighteen thousand specimens in his musem at Bologna. Bacon's proposal that the causes of oddities should be investigated was equally conventional. The depth of his thinking is, however, apparent when he turns to *why* we should concern ourselves with the causes of deformity. Bacon is not merely a physician with a physician's narrow interests. He is a philosopher with a philosopher's desire to know the nature of things. The critical passage is trenchant and lucid. We should, he says, study deviant instances '*For once a nature has been observed in its variations, and the reason for it has been made clear, it will be an easy matter to bring that nature by art to the point it reached by chance.*' Centuries ahead of his time Bacon recognised that the pursuit of the causes of error is not an end in itself, but rather just a means. The monstrous, the strange, the deviant, or merely the different, he is saying, reveal the laws of nature. And once we know those laws, we can reconstruct the world as we wish.

In a sense this book is an interim report on Bacon's project. It is not only about the human body as we might wish it to be, but as it is – replete with variety and error. Some of these varieties are the commonplace differences that give each of us our unique combinations of features and, as such, are a source of delight. Others are mere inconveniences that occupy the inter-tidal between the normal and the pathological. Yet others are the result of frank errors of development, that impair, sometimes grievously, the lives of those who have them, or simply kill them in early infancy. At the most extreme are deformities so acute that it is hardly possible to recognise those who bear them as being human at all.

Bacon's recommendation, that we should *collect* what he called 'prodigious births', may seem distasteful. Our ostensible, often ostentatious, love of human diversity tends to run dry when diversity shades into deformity. To seek out, look at, much less speak about deformity brings us uncomfortably close to naive, gaping wonder (or, to put it less charitably, prurience), callous derision, or at best a taste for thoughtless acquisition. It suggests the menageries of princes, the circuses of P.T. Barnum, Tod Browning's film *Freaks* (1932), or simply the basements of museums in which exhibits designed for our forebears' apparently coarser sensibilities now languish.

Yet the activity must not be confused with its objective. What were to Bacon 'monsters' and 'prodigious births' are to us just part of the spectrum of human form. In the last twenty years this spectrum has been sampled and studied as never before. Throughout the world, people with physiologies or physiognomies that are in some way or other unusual have been catalogued, photographed and pedigreed. They have been found in Botswana and Brazil, Baltimore and Berlin. Blood has been tapped from their veins and sent to laboratories for analysis. Their biographies, anonymous and reduced to the biological facts, fill scientific journals. They are, though they scarcely know it, the raw material for a vast biomedical enterprise, perhaps the greatest of our age, one in which tens of thousands of scientists are collectively engaged, and which has as its objective nothing less than the elucidation of the laws that make the human body.

Most of these people have mutations – that is, deficiencies in particular genes. Mutations arise from errors made by the machinery that copies or repairs DNA. At the time of writing mutations that cause some of us to look, feel, or behave differently from almost everyone else have been found in more than a thousand genes. Some of these mutations delete or add entire stretches of chromosome. Others affect only a single nucleotide, a single building block of DNA. The physical nature and extent of the mutation is not, however, as important as its consequences. Inherited disorders are caused by mutations that alter the gene's DNA sequence so that the protein it encodes takes a different, usually defective form, or simply isn't produced at all. Mutations alter the *meaning* of the genes.

Changing the meaning of a single gene can have extraordinarily far-flung effects on the genetic grammar of the body. There is a mutation that gives you red hair and also makes you fat. Another causes partial albinism, deafness, and fatal constipation. Yet another gives you short fingers and toes, and malformed genitals. In altering the meanings of genes, mutations give us a hint of what those genes meant to the body in the first place. They are collectively a Rosetta Stone that enables us to translate the hidden meanings of genes; they are virtual scalpels that slice through the genetic grammar and lay its logic bare.

Interpreting the meaning of mutations requires the adoption of a reverse logic that is, at first, counter-intuitive. If a mutation causes a child to be born with no arms, then, although it is tempting to speak of a gene for 'armlessness', such a mutation is really evidence for a gene that helps ensure that most of us do have arms. This is because most mutations destroy meaning. In the idiolect of genetics, they are 'loss-of-function' mutations. A minority of mutations add meaning and are called 'gain-of-function'. When interpreting the meaning of a mutation it is important to know which of these you are dealing with. One way to tell is by seeing how they are inherited. Loss-of-function mutations tend to be recessive: they will only affect a child's body when it inherits defective copies of the gene from both its parents. Gain-of-function mutations tend to be dominant: a child need have only one copy of the gene in order to see its effects. This is not an invariable distinction (some dominantly inherited mutations are loss-of-function) but it is a good initial guide. Gain or loss, both kinds of mutations reveal something about the function of the genes that they affect, and in doing so, reveal a small part of the genetic grammar. Mutations reverse-engineer the body.

Who, then, are the mutants? To say that the sequence of a particular gene shows a 'mutation', or to call the person who bears such a gene a 'mutant', is to make an invidious distinction. It is to imply, at the least, deviation from some ideal of perfection. Yet humans differ from each other in very many ways, and those differences are, at least in part, inherited. Who among us has the genome of genomes, the one by which all other genomes will be judged?

The short answer is that no one does. Certainly *the* human genome, the one whose sequence was published in *Nature* on 15 February 2001, is not a standard; it is merely a composite of the genomes of an unknown number of unknown people. As such, it has no special claim to normality or perfection (nor did the scientists who promoted and executed this great enterprise ever claim as much for it). This arbitrariness does not diminish in the slightest degree the value of this genomic sequence; after all, the genomes of any two people are 99.9 per cent identical, so anyone's sequence reveals almost everything about everyone's. On the other hand, a genome nearly three thousand million base-pairs long implies a few million base-pairs that differ between any two people; and it is in those differences that the interest lies.

If there is no such thing as a perfect or normal genome, can we find these qualities in a given gene? Perhaps. All of our thirty thousand genes show at least some variety. In the most recent generation of the world's inhabitants, each base-pair in the human genome mutated, on average, 240 times. Not all of these mutations change the meanings of genes or even strike genes at all. Some alter one of the vast tracts of the human genome that seem to be devoid of sense. Containing no genes that contribute to the grammar of the body, these regions are struck by mutation again and again; the scalpel slices but with no consequences to body or mind. Other mutations strike the coding regions of genes but do not materially alter the sequences of the proteins that they encode; these, too, are silent.

Of the mutations that alter the meaning of genes, a small minority will be beneficial and will become, with time, more common. So common, in fact, that it is hardly fair to refer to them as 'mutations', and instead we call them 'variants' or, more technically, 'polymorphisms'. In Africa, the $\Delta 32$ polymorphism of the CCR5 gene is currently increasing in frequency because it confers resistance to human immunodeficiency virus and so to AIDS. This is something new, but many polymorphisms are ancient. They are the stuff from which human diversity is made. They give us variety in skin colour, height, weight and facial features, and they surely also give us at least some of our variety in temperament, intelligence, addictive habits. They may cause disease, but mostly the diseases of old age such as senile dementia and heart attacks.

How common does a mutation have to be before it becomes a polymorphism? The answer is a bit arbitrary, but if a variant sequence has a global frequency of 1 per cent or more it is assumed that it cannot have caused much harm in its history, and may even have conferred some benefit to its carriers. By this criterion, at least one polymorphism has been detected in about 65 per cent of the human genes in which they have been sought, but some genes have dozens. This variety should not overwhelm us. Most human genes have one variant that is far more common than all others, and it is quite sensible to speak of that variant as being normal, albeit only in the statistical sense.

Perfection is far more problematic. The only reason to say that one genetic variant is 'better' than another is if it confers greater reproductive success on those who bear it; that is, if it has a higher Darwinian fitness than other variants. It is likely that the most common variant is the best under most circumstances, but this cannot be proved, for the frequencies of gene variants are shaped by history, and what was best then need not be best either now or in the future. To prefer one polymorphism over another – or rather to prefer the way it surfaces in our looks – is merely to express a taste. By this I mean the sort of claim made by the great French naturalist George Leclerc Buffon when he asserted that, for their fair skin and black eyes, the women of the Caucasus Mountains were lovelier than all others. Or when Karen Blixen eulogised the beauty of the Masai *morani*. Recognition of, even a delight in, human genetic diversity does not, however, commit us to a thorough-going genetic relativism. Many of the mutations that batter our genomes do us harm by any criterion.

Each new embryo has about a hundred mutations that its parents did not have. These new mutations are unique to a particular sperm or ovum, were acquired while these cells were in the parental gonads and were not present when the embryo's parents were themselves embryos. Of these hundred mutations, about four will alter the meaning of genes by changing the amino acid sequences of proteins. And of these four content-altering mutations, about three will be harmful. To be more precise, they will affect the ultimate reproductive success of the embryo, at least enough to ensure that, with time, natural selection will drive them to extinction.

These are uncertain numbers: the fraction of deleterious mutations can only be estimated by indirect methods. But if they are at all correct, their implications are terrifying. They tell us that our health and happiness are being continually eroded by an unceasing supply of genetic error. But matters are worse than that. Not only are we each burdened with our own unique suite of harmful mutations, we also have to cope with those we inherited from our parents, and they from theirs, and so on. What is the total mutational burden on the average human being? The length of time that a given mutation will be passed down from one generation to the next depends on the severity of its effects. If we suppose that an average mutation has only a mildly deleterious effect upon reproductive success and so persists for a hundred generations, an estimate of three new mutations per generation yields the depressing conclusion that the average newly conceived human bears three hundred mutations that impair its health in some fashion. No one completely escapes this mutational storm. But – and this is necessarily true – we are not all equally subject to its force. Some of us, by chance, are born with an unusually large number of mildly deleterious mutations, while others are born with rather few. And some of us, by chance, are born with just one mutation of devastating effect where most of us are not. Who, then, are the mutants? There can be only one answer, and it is one that is consistent with our everyday experience of the normal and the pathological. We are all mutants. But some of us are more mutant than others.

II

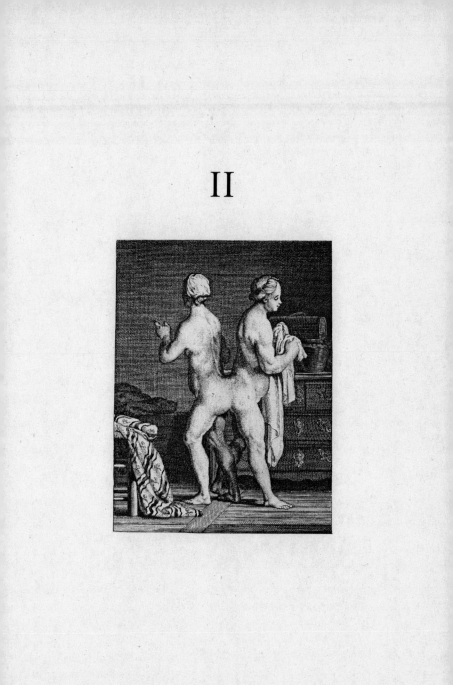

A PERFECT JOIN

[ON THE INVISIBLE GEOMETRY OF EMBRYOS]

IN THE VOLUME OF ENGRAVED PLATES that accompanies the report of their dissection, Ritta and Christina Parodi appear as a pair of small, slender, and quite beautiful infant girls. They have dark eyes, and their silky curls are brushed forward over their foreheads in the fashion of the French Empire, in a way that suggests a heroic portrait of Napoleon Bonaparte. Their brows and noses are straight, their mouths sweetly formed, and their arms reach towards each other, as if in embrace, but their expressions are conventionally grave. Distinct from the shoulders up, their torsos melt gradually into each other; below the single navel the join is so complete that they have, between them, one vulva, one rectum, one pelvis, and one pair of legs. It is a paradoxical geometry. For although the girls

CONJOINED TWINS: PYGOPAGUS. JUDITH AND HÉLÈNE (1701–23). FROM GEORGE LECLERC BUFFON 1777 *Histoire naturelle générale et particulière.*

are, individually, so profoundly deformed, together they are symmetrical and proportionate; their construction seems less an anomaly of nature than its designed result. It may be thought that this beauty is merely a product of the engraver's art, but a plaster-cast of their body shows the same harmony of form. If the engraver erred it was only in giving them the proportions of children older than they were; they were only eight months old when they died.

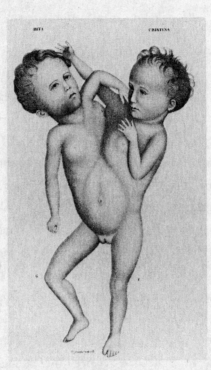

CONJOINED TWINS: PARAPAGUS DICEPHALUS TETRABRACHIUS. RITTA AND CHRISTINA PARODI (1829). FROM ÉTIENNE SERRES 1832 *Recherches d'anatomie transcendante et pathologique.*

THE APOTHEOSIS OF RITTA-CHRISTINA

The Parodis arrived in Paris in the autumn of 1829. Six months previously they had left Sassari, a provincial Sardinian town, in the hope of living by the exhibition of their children. Italy had been receptive; Paris was not. Local magistrates, ruling on the side of public decency, forbade the Parodis to show their children to the multitude and so deprived them of their only income. They moved to a derelict house on the outskirts of the city, where they received some payment from a procession of physicians and philosophers who came to see the children in private.

What they earned wasn't even enough to heat the house. The *savants*, puzzling over what they found, were also continually uncovering the children. Was there one heart or two? The stethoscope gave conflicting results. They were fascinated by the differences between the children. Christina was a delight – healthy, vigorous, with a voracious appetite; Ritta, by contrast, was weak, querulous and cyanotic. When one fell asleep the other would usually do so as well, but occasionally one slept soundly while the other demanded food. Continually exposed to chills, Ritta became bronchitic. The physicians noted that sickness, too, demonstrated the dual and yet intertwined nature of the girls, for even as Ritta gasped for air, her sister lay at her side unaffected and content. But three minutes after Ritta died, Christina gave a cry and her hand, which was in her mother's, went limp. It was 23 November 1829, and the afterlife of 'Ritta-Christina, the two-headed girl' had begun.

* * *

The men from the Académie Royale de Médecine were on hand within hours. They wanted a cast of the body. Deputations of anatomists followed; they wanted the body itself. How they got it is a murky affair, but within days the dissection of *l'enfant bicéphale* was announced. In the vast amphitheatre of the Muséum d'Histoire Naturelle at the Jardin des Plantes in Paris, Ritta and Christina were laid out in state on a wooden trestle table. The anatomists jostled for space around them. Baron Georges Cuvier, France's greatest anatomist – 'the French Aristotle' – was there. So was Isidore Geoffroy Saint-Hilaire, connoisseur of abnormality, who in a few years would lay the foundation of teratology. And then there was Étienne Reynaud Augustin Serres, the brilliant young physician from the Hôpital de la Pitié, who would make his reputation by anatomising the girls in a three-hundred-page monograph.

Beyond the walls of the museum, Paris was enthralled. The *Courier Français* intimated that the medical men had connived at the death of the sisters; they replied that the magistrates who had let the family sink to such miserable depths were to blame. The journalist and critic Jules Janin published a three-thousand-word *j'accuse* in which he excoriated the anatomists for taking the scalpel to the poetic mystery that was Ritta and Christina: 'You despoil this beautiful corpse, you bring this monster to the level of ordinary men, and when all is done, you have only the shade of a corpse.' And then he suggested that the girls would be a fine subject for a novel.

The first cut exposed the ribcage. United by a single sternum, the ribs embraced both sisters, yet were attached to two quite distinct vertebral columns that curved gracefully down to the common pelvis. There were two hearts, but they were contained within a single pericardium, and Ritta's was profoundly deformed: the intra-auricular valves were perforated and she had two superior vena cavas, one of which opened into the left ventricle, the other into the right – the likely cause of her cyanosis. Had it not been for this imperfection, lamented Serres, and had the children lived under more favourable circumstances, they would surely have survived to adulthood. Two oesophagi led to two stomachs, and two colons, which then joined to a common rectum. Each child had a uterus, ovaries and fallopian tubes, but only one set of reproductive organs was connected to the vagina, the other being small and underdeveloped. Most remarkably of all, where Christina's heart, stomach and liver were quite normally oriented, Ritta's were transposed relative to her sister's, so that the viscera of the two girls formed mirror-images of each other. The anatomists finished their work, and then boiled the skeleton for display.

A PAIR OF LONG-CASE CLOCKS

The oldest known depiction of a pair of conjoined twins is a statue excavated from a Neolithic shrine in Anatolia. Carved from white marble, it depicts a pair of dumpy middle-aged women joined at the hip. Three thousand years after this statue was carved, Australian Aborigines inscribed a memorial to a dicephalus (two heads, one body) conjoined twin on a rock that lies near what are now the outskirts of Sydney. Another two thousand years (we are now at 700 BC), and the conjoined Molionides brothers appear in Greek geometric art. Eurytos and Cteatos by name, one is said to be the son of a god, Poseidon, the other of a mortal, King Actor. Discordant paternity notwithstanding, they have a common trunk and four arms, each of which brandishes a spear. In a Kentish parish, loaves of bread in the shape of two women locked together side by side are distributed to the poor every Easter Monday, a tradition, it is said, that dates from around the time of the Norman conquest and that commemorates a bequest made by a pair of conjoined twins who once lived there.

By the sixteenth century, conjoined twins crop up in the monster-and-marvel anthologies with the monotonous regularity with which they now appear in British tabloids or the *New York Post*. Ambroise Paré described no fewer than thirteen, among them two girls joined back to back, two sisters joined at the forehead, two boys who shared a head and two infants who shared a heart. In 1560 Pierre Boaistuau gave an illuminated manuscript of his *Histoires prodigieuses* to Elizabeth I of England. Amid the plates of demonic creatures, wild men and fallen monarchs, is one devoted to two young women standing in a field on a single pair of legs, flaming red hair falling over their shoulders, looking very much like a pair of Botticelli Venuses who have somehow become entangled in each other.

For the allegory-mongers, conjoined twins signified political union. Boaistuau notes that another pair of Italian conjoined twins were born on the very day that the warring city-states of Genoa and Venice had finally declared a truce – no coincidence

CONJOINED TWINS: PARAPAGUS DICEPHALUS DIBRACHIUS.
NORMANDY. FROM PIERRE BOAISTUAU 1560
Histoires prodigieuses.

there. Montaigne, however, will have none of it. In his *Essays*
(c.1580) he describes a pair of conjoined twins that he encountered as they were being carted about the French countryside by
their parents. He considers the idea that the children's joined
torsos and multiple limbs might be a comment on the ability of
the King to unify the various factions of his realm under the rule
of law, but then rejects it. He continues, *'Those whom we call
monsters are not so with God, who in the immensity of his work seeth
the infinite forms therein contained.'* Conjoined twins did not
reflect God's opinion about the course of earthly affairs. They
were signs of His omnipotence.

By the early eighteenth century, this humanist impulse – the same impulse that caused Sir Thomas Browne to write so tenderly about deformity – had arrived at its logical conclusion. In 1706 Joseph-Guichard Duverney, surgeon and anatomist at the Jardin du Roi in Paris, the very place where Ritta and Christina had been laid open, dissected another pair of twins who were joined at the hips. Impressed by the perfection of the join, Duverney concluded that they were without doubt a testament to the 'the richness of the Mechanics of the Creator', who had clearly designed them so. After all, since God was responsible for the form of the embryo, He must also be responsible if it all went wrong. Indeed, deformed infants were not really the result of embryos gone *wrong* – they were part of His plan. Bodies, said Duverney, were like clocks. To suppose that conjoined twins could fit together so nicely without God's intervention was as absurd as supposing that you could take two long-case clocks, crash them into each other, and expect their parts to fuse into one harmonious and working whole.

Others thought this was ridiculous. To be sure, they argued, God was ultimately responsible for the order of nature, but the notion that He had deliberately engineered defective eggs or sperm as a sort of creative flourish was absurd. If bodies were clocks, then there seemed to be a lot of clocks around that were hardly to the Clockmaker's credit. Monsters were not evidence of divine design: they were just accidents.

The conflict between these two radically different postitions, between deformity as divine design and deformity as accident, came to be known as *la querelle des monstres* – the quarrel of the monsters. It pitted French anatomists against one another for decades, the contenders trading blows in the *Mémoires de l'Académie Royale des Sciences*. More than theology was at stake. The quarrel was also a contest over two different views of how embryos are formed. Duverney and his followers were pre-formationists. They held that each egg (or, in some version of the theory, each sperm) contained the entire embryo writ small, complete with limbs, liver and lungs. Stranger yet, this tiny embryo (which some microscopists claimed they could see) also contained eggs or sperm, each of which, in turn contained an embryo...and so on, *ad infinitum*. Each of Eve's ovaries, by this reasoning, contained all future humanity.

Preformationism was an ingenious theory and won promi-nent adherents. Yet many seventeenth- and eighteenth-century philosophers, among them freethinkers such as Buffon and Maupertuis, preferred some version of the older theory of 'epigenesis', the notion that embryonic order does not exist in the egg or the sperm *per se*, but rather emerges spontaneously after fertilisation. At the time of the *querelle*, many thought that the preformationists had the better side of the argument. Today, however, it is more difficult to judge a victor. Neither the preformationists nor the epigeneticists had a coherent theory of inheritance, so the terms of the debate between them do not correspond in any simple way to a modern understanding of the causes of deformity or development. Preformationism, with its infinite regress of embryos, seems the more outlandish of the two theories, though it captures nicely the notion that develop-ment errors are often (though not invariably) due to some mistake intrinsic to the germ cells – the cells that become eggs and sperm – or at least their DNA. But the epigeneticists speak more powerfully to the idea that embryos are engaged in an act of self-creation which can be derailed by external influences, chemicals and the like, or even chance events within their dividing cells.

HOW TO MAKE A CONJOINED TWIN

What makes twins conjoin? Aristotle, characteristically, covered the basic options. In one passage of *The generation of animals* he argues that conjoined twins come from two embryos that have fused. That, at least, is where he thought conjoined chickens (which have four wings and four legs) come from. But elsewhere he suggests that they come from one embryo that has split into two.

To modern ears his notion of how an embryo might split sounds odd, but it is a sophisticated account, all of a piece with his theory of how embryos develop. Having no microscope, Aristotle knows nothing of the existence of sperm and eggs. Instead he supposes that embryos coagulate out of a mixture of menstrual fluid and semen, the semen causing the menstrual fluid to thicken rather as – to use his homely metaphor – fig juice causes milk to curdle when one makes cheese. This is epigenesis *avant la lettre*. Indeed, preformationism was very much an attack on the Aristotelian theory of embryogenesis and, by extension, its account of the origins of deformity. Sometimes, says Aristotle, there is simply too much of the pre-embryonic mix. If there is only a little too much, you get infants with extra or unusually large parts, such as six fingers or an overdeveloped leg; more again, and you get conjoined twins; even more mix, separate twins. He uses a beautiful image to describe how the mix separates to make two individuals. They are, he says, the result of a force in the womb like falling water: '...as the water in rivers is carried along with a certain motion, if it dash against anything two systems come into being out of one, each retaining the same motion; the same thing happens with the embryo'.

For Aristotle, the two ways of making conjoined twins bear on their individuality. He rules that if conjoined twins have separate hearts, then they are the products of two embryos and are two individuals; if there is only one heart, then they are one. The question of conjoined twin individuality haunts their history. Thomas Aquinas thought that it depended on the number of hearts *and* heads (thereby ensuring perpetual confusion for priests who wanted to know how many baptisms conjoined infants required). When twins are united by only by a slender cartilaginous band – the case with the original Siamese twins, Eng and Chang (1811–74) – it is easy to grant each his own identity. More intimately joined twins have, however, always caused confusion. In accounts of Ritta and Christina Parodi, the girls often appear as the singular 'Ritta-Christina', or even 'the girl with two heads', rather than two girls with one body – which is what they were.

Until recently, the origin of conjoined twins has been debated in much the terms that Aristotle used: they are the result either of fusion or fission. Most medical textbooks plump for the latter. Monozygotic (identical) twins, the argument goes, are manifestly the products of one embryo that has accidentally split into two; and if an embryo can split completely, surely it can split partially as well. This argument has the attraction of simplicity. It is also true that conjoined twins are nearly always monozygotic – they originate from a single egg fertilised by a single sperm. Yet there are several hints that monozygotic twins who are born conjoined are the result of quite different events in the first few weeks after conception than are those who are born separate.

One difference between conjoined and separate twins is that conjoined twins share a single placenta and (as they must) a single amniotic sac. Separate twins also share a single placenta, but each usually has an amniotic sac of its own as well. Since the amniotic sac forms after the placenta, this suggests that the split – if split it is – happens later in conjoined twins than in separate twins.

Another suggestive difference comes from the strange statistics of twin gender. Fifty per cent of separate monozygotic twins born are female. This is a little higher than one would expect, since, in most populations at most times, slightly fewer girls than boys are born. But in conjoined twins the skew towards femininity is overwhelming: about 77 per cent are girls. No one knows why this is so, but it neatly explains why depictions of conjoined twins – from Neolithic shrines to the *New York Post* – are so often female.

Perhaps the best reason for thinking that conjoined twins are not the result of a partially split embryo is the geometry of the twins themselves. Conjoined twins may be joined at their heads, thoraxes, abdomens or hips; they may be oriented belly to belly, side to side, or back to back; and each of these connections may be so weak that they share hardly any organs or so intimate that they share them all. It is hard to see how all this astonishing array of bodily configurations could arise by simply splitting an embryo in two.

But where are the origins of conjoined twins to be found if not in partially split embryos? Sir Thomas Browne called the womb 'the obscure world', and so it is – never more so than when we try to explain the creation of conjoined twins. The latest ideas suggest, however, that Aristotle's dichotomy – fission *or* fusion – is illusory. The making of conjoined twins is, first, a matter of making two embryos out of one, and then of gluing them together. Moreover, the way in which two embryos are made out of one is nothing so crude as some sort of mechanical splitting of the embryo. It is, instead, something more subtle and interesting. Indeed, although we perceive conjoined twins as the strangest of all forms that the human body can take (as recently as 1996 *The Times* referred to one pair of twins as 'metaphysical insults'), they have shown us the devices by which our bodies are given order in the womb.

ORGANISE ME

On the seventh day after conception, a human embryo begins to dig. Though only a hollow ball made up of a hundred or so cells, it is able to embed itself in the uterine linings of its mother's womb that are softened and swollen by the hormones of the menstrual cycle. Most of the cells in the hollow ball are occupied with the business of burrowing, but some are up to other things. They are beginning to organise themselves into a ball of their own, so that by day 9 the embryo is rather like one of those ingenious Chinese toys composed of carved ivory spheres within spheres within spheres. By day 13 it has disappeared within the uterine lining, and the wound it has caused has usually healed. The embryo is beginning to build itself.

Its first task is to make the raw materials of its organs. We are three-dimensional creatures: bags of skin that surround layers of bone and muscle that, in turn, support a maze of internal plumbing; and each of these layers is constructed from specialised tissues. But the embryo faces a problem. Of the elaborate structure that it has already built, only a minute fraction – a small clump of cells in the innermost sphere – is actually destined to produce the foetus; all the rest will just become its ancillary equipment: placenta, umbilical cord and the like. And to make foetus out of this clump of cells, the embryo has to reorganise itself.

The process by which it does this is called 'gastrulation'. At about day 13 after conception, the clump of cells has become a disc with a cavity above it (the future amniotic cavity) and a cavity below it (the future yolk sac). Halfway down the length of this disc, a groove appears, the so-called 'primitive streak'. Cells migrate towards the streak and pour themselves into it. The first cells that go through layer themselves around the yolk cavity. More cells enter the streak and form another layer above the first. The result is an embryo organised into three layers where once there was one: a gastrula.

The three layers of the gastrula anticipate our organs. The top layer is the ectoderm – it will become the outer layers of the skin and most of the nervous system; beneath it is the mesoderm – future muscle and bone; and surrounding the yolk is the endoderm – ultimate source of the gut, pancreas, spleen and liver. (*Ecto-*, *meso-* and *endo-* come from the Greek for outer, middle and inner *derm* – skin – respectively.)

The division sounds clear-cut, but in fact many parts of our bodies – teeth, breasts, arms, legs, genitalia – are intricate combinations of ectoderm and mesoderm. More important than the material from which it builds its organs, the embryo has also now acquired the geometry that it will have for the rest of its life. Two weeks after egg met sperm, the embryo has a head and a tail, a front and a back, and a left and a right. The question is, how did it get them?

In the spring of 1920, Hilda Pröscholdt arrived in the German university town of Freiburg. She had come to work with Hans Spemann, one of the most important figures in the new, largely German, science of *Entwicklungsmechanik*, 'developmental mechanics'. The glassy embryos of sea urchins were being bisected; green-tentacled *Hydra* lost their heads only to regrow them again; frogs and newts were made to yield up their eggs for intricate transplantation experiments. Spemann was a master of this science, and Pröscholdt was there to do a Ph.D. in his laboratory. At first she floundered; the experiments that Spemann asked her to do seemed technically impossible and, in retrospect, they were. But she was bright, tenacious and competent, and in the spring of 1921 Spemann suggested another line of work. Its results would provide the first glimpse into how the embryo gets its order.

Then as now, the implicit goal of most developmental biologists was to understand how human embryos construct themselves, or failing that, how the embryos of other species of mammal do. But mammal embryos are difficult to work with. They're hard to find and difficult to keep alive outside the womb. Not so newt embryos. Newts lay an abundance of tiny eggs that can, with practice, be surgically manipulated. It was even possible to transplant pieces of tissue between newt embryos and have them graft and grow.

The experiment which Spemann now suggested to Hilda Pröscholdt entailed excising a piece of tissue from the far edge of one embryo's blastopore – the newt equivalent of the human primitive streak – and transplanting it onto another embryo. Observing that the embryo's tissue layers and geometry arose from cells that had passed through the blastopore, Spemann reasoned that the tissues at the blastopore's lip had some special power to instruct the cells that were travelling past it. If so, then embryos that had extra bits of blastopore lip grafted onto them might have – what? Surplus quantities of mesoderm and endoderm? A fatally scrambled geometry? Completely normal development? Earlier experiments that Spemann himself had carried out had yielded intriguing but ambiguous results. Now Hilda Pröscholdt was going to do the thing properly.

Between 1921 and 1923 she carried out 259 transplantation experiments. Most of her embryos did not survive the surgery. But six embryos that did make it are among the most famous in developmental biology, for each contained the makings of not one newt but two. Each had the beginnings of two heads, two tails, two neural tubes, two sets of muscles, two notochords, and two guts. She had made conjoined-twin newts, oriented belly to belly.

This was remarkable, but the real beauty of the experiment lay in Pröscholdt's use of two different species of newts as donor and host. The common newt, the donor species, has darkly pigmented cells where the great-crested newt, the host species, does not. The extra organs, it was clear, belonged to the host embryo rather than the donor. This implied that the transplanted piece of blastopore lip had not *become* an extra newt, but rather had *induced* one out of undifferentiated host cells. This tiny piece of tissue seemed to have the power to instruct a whole new creature, complete in nearly all its parts. Spemann, with no sense of hyperbole, called the far lip of the newt's blastopore 'the organiser', the name by which it is still known.

For seventy years, developmental biologists searched in vain for the source of the organiser's power. They knew roughly what they were looking for: a molecule secreted by one cell that would tell another cell what to do, what to become, and where to go.

Very quickly it became apparent that the potency of the organiser lay in a small part of mesoderm just underneath the lip of the blastopore. The idea was simple: the cells that had migrated through the blastopore into the interior of the embryo were naive, uninformed, but their potential was unlimited. Spemann aphorised this idea when he said 'We are standing and walking with parts of our body that could have been used for thinking had they developed in another part of the embryo.' The mesodermal cells of the blastopore edge were the source of a signal that filtered into the embryo, or to use the term that was soon invented, a *morphogen*. This signal was strong near its source but gradually became fainter and fainter as it dissipated away. There was, in short, a three-dimensional gradient in the concentration of morphogen. Cells perceived this gradient and knew accordingly where and what they were. If the signal was strong, then ectodermal cells formed into the spinal cord that runs the length of our back; if it was faint, then they became the skin that covers our body. The same logic applied to the other germ layers. If the organiser signal was strong, mesoderm would become muscle; fainter, kidneys; fainter yet, connective tissue and blood cells. What the organiser did was pattern the cells beneath it.

It would be tedious to recount the many false starts, the years wasted on the search for the organiser morphogen, the hecatombs of frog and newt embryos ground up in the search for the elusive substance, and then, in the 1960s, the growing belief that the problem was intractable and should simply be abandoned. 'Science,' Peter Medawar once said, 'is the Art of the Soluble.' But the soluble was precisely what the art of the day could not find.

In the early 1990s recombinant DNA technology was applied to the problem. By 1993 a protein was identified that, when injected into the embryos of African clawed toads, gave conjoined-twin tadpoles. At last it was possible to obtain – without crude surgery – the results that Hilda Pröscholdt had found so many years before. The protein was especially good at turning naive ectoderm into spinal cord and brain. With a whimsy that is pervasive in this area of biology, it was named 'noggin'. By this time techniques had been developed that made it possible to see where in an embryo genes were being switched on and off. The noggin gene was turned on at the far end of the blastopore's lip, just where the gene encoding an organising morphogen should be.

Noggin is a signalling molecule – that is, a molecule by which one cell communicates with another. Animals have an inordinate number of them. Of the thirty thousand genes in the human genome, at least twelve hundred are thought to encode proteins involved in communication between cells. They come in great families of related molecules: the transforming growth factor-betas (TGF-β), the hedgehogs and the fibroblast growth factors (FGFs) to name but a few, and some families contain more than a dozen members. The way they work varies in detail, but the theme is the same. Secreted by one cell, they attach to receptors on the surfaces of other cells and in doing so begin a sequence of molecular events that reaches into the recipient cell. The chain of information finally reaches the nucleus, where batteries of other genes are either activated or repressed, and the cell adopts a fate, an identity.

When noggin was first discovered, it was supposed that its uncanny powers lay in an ability to define the back of the embryo from the front – more precisely, to instruct naive ectodermal cells to become spinal column rather than skin. This was the simplest interpretation of the data. Noggin, the thinking went, spurred ectodermal cells on to higher things; without it, they would languish as humble skin.

The truth is a bit more subtle. The probability that a cell becomes spinal column rather than skin is not just a function of the quantity of noggin that finds its way to its receptors, but is rather the outcome of molecular conflict over its fate. I said that our genomes encode an inordinate number of signalling molecules. This implies that the cells in our bodies must be continually bathed in many signals emanating from many sources. Some of these signals speak with one voice, but others offer conflicting advice. Noggin from the organiser may urge ectoderm to become neurons, but as it does so, from the opposite side of the embryo another molecule, bone morphogenetic protein 4 (BMP4) instructs those same cells to become skin.

The manner in which the embryo resolves the conflict between these two signals is ingenious. Each signal has its own receptor to which it will attach, but noggin, with cunning versatility, can also attach to free BMP4 molecules as they filter through the intercellular spaces, and disable them. Cells close to the organiser are not only induced to become neurons, but are also inhibited from becoming skin; far from the organiser the opposite obtains. The fate of a given cell depends on the balance of the concentration between the two competing molecules. It is an ingenious device, only one of many like it that work throughout the development of vertebrate bodies, at scales large and small, to a variety of ends; but here the end is a toad or a child that has a front and a back. In a way, the embryo is just a microcosm of the cognitive world that we inhabit, the world of signals that insistently urge us to travel to one destination rather than another, eschew some goals in favour of others, hold some things to be true and others false; in short, that moulds us into what we are.

It is actually quite hard to prove that a gene, or the protein that it encodes, does what one supposes. One way of doing so is to eliminate the gene and watch what happens. This is rather like removing a car part – some inconspicuous screw – in order to see why it's there. Sometimes only a rear-view mirror falls off, but sometimes the car dies. So it is with mice and genes. If noggin were indeed the long-sought organising molecule, then any mouse with a defective noggin gene should have a deeply disordered geometry. For want of information, the cells in such an embryo would not know where they were or what to do. One might expect a mouse that grew up in the absence of noggin to have no spinal column or brain, but be belly all round; at the very least one would expect it to die long before it was born. Oddly enough, when a noggin-defective mouse was engineered in 1998, it proved to be really quite healthy. True, its spinal cord and some of its muscles were abnormal, but its deformities were trivial compared to what they might have been.

The reason for this is still not completely understood, but it probably lies in the complexity of the organiser. Since the discovery of noggin at least seven different signalling proteins have been found there, among them the ominously named 'cerberus' (after the three-headed dog that guards the entrance to Hades), and the blunter but no less evocative 'dickkopf' (German for 'fat-head'). This multiplicity is puzzling. Some of these proteins probably have unique tasks (perhaps giving pattern to the head but not the tail, or else ectoderm but not mesoderm), but it could also be that some can substitute for others. Biologists refer to genes that perform the same task as others as 'redundant' in much the same sense that employers do: one can be disposed of without the enterprise suffering ill-effects. At least two of the organiser signals, noggin and another called chordin, appear to be partially redundant. Like noggin, chordin instructs cells to become back rather than front, neurons rather than skin, and does so by inhibiting the BMP4 that filters up from the opposite side of the embryo. And, like noggin-defective mice, mice engineered with a defect in the chordin gene have more or less normal geometry, although they are stillborn. However, doubly-mutant mice, in which both the noggin and chordin genes have been disabled, never see the light of day. The doubly-mutant embryos die long before they are born, their geometries profoundly disordered. They can only be found by dissecting the mother in early pregnancy.

Hilda Pröscholdt's results were published in 1924, but she did not live to see them in print. Halfway through her doctoral degree she married Otto Mangold, one of her fellow students in Spemann's laboratory, and it is by his name that she is now known. In December 1923, having been awarded a doctorate, she gave birth to a son, Christian, and left the laboratory. On 4 September 1924, while visiting her Swabian in-laws, she spilt kerosene while refuelling a stove. Her dress caught alight, and she died the following day of her burns. She was only twenty-six, and in all ways a product of the Weimar. As a student, when not dissecting embryos, she had read Rilke and Stefan George, sat in on the philosopher Edmund Husserl's lectures, decorated her flat with Expressionist prints, and taken long Black Forest walks. She had only really done one good experiment, but it is said by some that had Hilda Pröscholdt lived she would have shared the Nobel Prize that Spemann won in 1935.

E PLURIBUS UNUM?

When Eng and Chang toured the United States they advertised themselves with the slogan, familiar to any citizen of the Republic, *e pluribus unum* – out of many, one. It seemed apt enough, but it was only half the truth. Conjoined twins are clearly, in the first instance, a case of *ex uno plures* – out of one, many.

The similarity of human twins to the conjoined-twin newts made by Hilda Pröscholdt suggests one way how this might happen. All that is needed are two organisers on a single embryo instead of the usual one. Although Pröscholdt doubled the organisers on her newts by some deft, if crude, transplantation surgery, there are much more subtle molecular means of bringing about the same end. The genes that encode the signalling proteins of the organiser – noggin, cerberus, dickkopf and so on – are regulated by yet other 'master control genes'. The making of two embryos out of one may, therefore, be simply a matter of one of these master control genes being turned on in the embryo where it normally is not. Why this should happen is a mystery – human conjoined twins occur so rarely (about 1 in every 100,000 live births) and unpredictably that there is no obvious way to

CONJOINED TWINS: PARAPAGUS DICEPHALUS DIBRACHIUS.
FROM B.C. HIRST AND G.A. PIERSOL 1893
Human monstrosities.

find out. Perhaps they are caused by chemicals in the environment: at least one drug (albeit a rare and potent chemothera-
peutic agent) has been shown to cause conjoined twinning in
mice. Whatever the ultimate cause of conjoined twins, the 'two-
organiser' theory, while a neatly plausible account of how to get
two embryos out of one, is not in itself a complete explanation
for their existence. The theory has nothing to say about their
essential feature: the fact that they are glued together.

One man who thought deeply about the conjoinedness of conjoined twins was Étienne Geoffroy Saint-Hilaire. In 1829
Geoffroy was Professor at the Muséum d'Histoire Naturelle,
and next to Cuvier (his colleague and bitter rival) the most

important anatomist in France. Geoffroy's disciple Étienne Serres had written the monograph describing Ritta and Christina Parodi's autopsy; Geoffroy's son, Isidore, had organised the event. It is upon Isidore that suspicion falls for having bullied the Parodis into surrendering the corpse.

Geoffroy *père* was one of the most mercurial intellects of his time: almost everything he wrote has a touch of genius and a touch of the absurd. He was one of nature's romantics: ostensibly a descriptive anatomist, he investigated the devices by which puffer fish inflate themselves, but did not shy away from larger problems, such as the relationships between the 'imponderable fluids' of the universe (light, electricity, nervous energy, etc.), his deductive theory of which never saw print. More reasonably, Geoffroy was also keenly interested in deformity. It is in his hands that teratology first really becomes a science.

In 1799 Geoffroy was among the *savants* that Napoleon Bonaparte brought to Egypt in his futile attempt to block England's route to the East. Geoffroy spent his Egyptian sojourn (cut short by the arrival of the British) collecting crocodiles, ichneumons and mummified ibises. Egypt also gave him a way of making 'monsters' to order. Geoffroy was a staunch epigeneticist. If monsters were caused by accidents in the womb, he reasoned, it should be possible to engineer them. Since time immemorial, the peasants of the Nile valley had incubated chicken eggs in earthenware furnaces fired by burning cowdung. Inspired by this, Geoffroy established a similar hatchery where he systematically abused developing eggs by shaking them around, perforating them, or covering them in gold foil.

The resulting chicks were mostly more dead than deformed, but some had bent digits, odd-looking beaks and skulls, and a few lacked eyes – unspectacular results, but enough to convince Geoffroy that he had definitively slain preformationism.

From monstrous chickens to monstrous humans was an easy leap and, starting in 1822, Geoffroy published a string of papers on deformed infants, which he classified as zoologists classify insects. A child whose head was externally invisible belonged, for example, to the genus *Cryptocephalus*. He realised that his 'genera' were not specific to humans: dogs, cats, perhaps even fish, could be deformed in the same way; his classification transcended the scale of nature. A few years later Isidore elaborated his father's classification into a system that is still, with some modification, used by teratologists today, one in which Ritta and Christina, and children like them, are known as 'Xiphopages' to the French and 'parapagus dicephalus tetrabrachius' (side-joined, two-headed four-armed) conjoined twins to everyone else.

Étienne Geoffroy Saint-Hilaire's greatest contribution to teratology was, however, the realisation that deformity is a natural consequence of the laws that regulate the development of the human body. Moreover, looked at the right way, such deformed infants can *reveal* those laws. This, of course, is a very Baconian idea – and in one of his more philosophical tracts the anatomist speaks warmly of the genius of James I's Lord Chancellor.

Nowhere, for Geoffroy, were those laws more clearly revealed than in conjoined twins. Even before seeing Ritta and Christina Parodi in 1829, he had dissected a number of conjoined twins. Conjoinedness, he argued, was simply a reflection of what normally happens in a single embryo. The organs of an embryo develop from disparate parts that are then attracted to each other by a mysterious force rather like gravity. The intimacy of conjoined twins is caused by this same force, but misapplied so that the parts of neighbouring embryos fuse instead to one another.

Geoffroy was deeply enamoured of this deduction and, in the positivist fashion of his day, made a law of it: *le loi d'affinité de soi pour soi* – the law of affinity of like for like. In the monograph that Étienne Serres wrote on Ritta and Christina's dissection, fully the first half is devoted to the *soi pour soi* and a few other laws of Serres's own devising. Geoffroy regarded the *soi pour soi* as his greatest discovery, and in later years elevated it into a fundamental law of the universe, not unlike Goethe's notion of 'elective affinities' to which it is related. This hubristic vision has ensured that the *soi pour soi* is, today, quite forgotten. This is a pity, since although Geoffroy's law is unsatisfactorily vague, and wrong in detail, it conveys something important about how human embryos are built. It was the first scientific explanation of connectedness.

CONNECTEDNESS

Eighteen days after conception the embryo is just a white, oval disc about a millimetre long. It has no organs, just three tissue layers and a geometry. Even the geometry is largely virtual: a matter of molecules that have been ordered in space and time, but not yet translated into anything that can be seen without the special stains that molecular biologists use. Within the next ten days all this will have changed. The embryo will be recognisably an incipient human – or at least some sort of vertebrate, a dog, a chicken or perhaps a newt. It will have a head, a neck, a spinal column, a gut; it will have a heart.

The first sign of all this future complexity comes on day 19 when a sheet of tissue, somewhat resembling the elongate leaf of a tulip, forms down the middle of the embryo above the primitive streak. The leaf isn't entirely flat: its edges show a tendency to furl to the middle, so that if you were to make a transverse section through the embryo you would see that it forms a shallow U. By the next day the U has become acute. Two more days and its vertices have met and touched in the middle of the embryo, rather as a moth folds its wings. And then the whole thing zips up, so that by day 23 the embryo has a hollow tube that runs most of its length, the nature of which is now clear: it is the beginnings of the mighty tract of nerves that we know as the spinal column. At one end, you can even see the rudiments of a brain.

Even as the nerve cord is forming, the foundations of other organs are being laid. Small brick-like blocks of tissue appear either side of incipient nerve cord, at first just a few, but then ten, twenty, and finally forty-four. Made of mesoderm, they reach around the neural tube to meet their opposite numbers and encase the neural tube. They will become vertebrae and muscles and the deepest layers of the skin. Underneath the embryo the endoderm, which embraces an enormous, flaccid sac of yolk, retracts up into the embryo to become the gut. As the gut shrinks the two halves of the embryo that it has previously divided are drawn together. Two hitherto inconspicuous tubes, one on either side, then unite to make a single larger tube running the length of the embryo's future abdomen, an abdominal tube that echoes the neural tube on its back. Within a few days this abdominal tube will begin to twist and then twist again to become a small machine of exquisite design. Though it still looks nothing like what it will become it already shows the qualities that led William Harvey to call it 'the Foundation of Life, the Prince of All, the Sun of the Microcosm, on which all vegetation doth depend, from whence all Vigor and Strength doth flow'. On day 21 it begins to beat.

The ability of disparate organ primordia to find each other and fuse to form wholes is one of the marvels of embryogenesis. Underlying it are thousands of different molecules that are attached to the surface of cells and are, as it were, signals of their affiliation, that permit other cells to recognise them as being of like kind. These are the cell-adhesion molecules; molecular biologists speak of them as the Velcro of the body: weak individually, but collectively strong. Even so, the fusion of organ primordia is a delicate business. Neural tube fusion is particularly prone to failure. One infant in a thousand born has a neural tube that is at least partly open – a condition called spina bifida. At its most severe the neural tube in the future head fails to close. The exposed neural tissue becomes necrotic and collapses, leaving a child that has the remnant of a brain stem but in which the back of the head has been truncated, as if sliced with a cleaver.

Such anencephalic infants, as they are known, occur in about 1 in 1500 births; they have heavy-lidded eyes that seem to bulge from their heads and their tongues stick out of their mouths. They die within a few days, if not hours, of being born. As the name suggests, spina bifida is often not just a failure of the neural tube to close, but a failure in the closure of the vertebral column so that instead of being sheltered by bone the nerve cord lies exposed. It is not the only organ prone to this sort of defect. Sometimes the primordia of the heart fail to meet; the result is cardiac bifida, two hearts, each only half of what it should be.

The power of cell–cell adhesion to mould the developing body is startling. In his monograph on Ritta and Christina, Serres describes a pair of stillborn boys who are joined at the head. Oriented belly to belly, their faces are deflected ninety degrees relative to their torsos so that they gaze, Janus-like, in opposite directions. What is remarkable about these children is that each apparent face is composed of half of one child's face

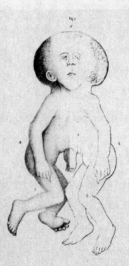

CONJOINED TWINS: CEPHALOTHORACOILEOPAGUS.
FROM ÉTIENNE SERRES 1832 *Recherches d'anatomie transcendante et pathologique.*

fused to the opposite half of his brother's. The developing noses, lips, jaws and brains of these two children have found each other and fused perfectly – twice.

The diversity of ways in which conjoined twins can be attached to each other seems to depend on the position of the developing embryonic discs relative to each other as they float on their common yolk sac and when they contact. The embryonic discs that gave rise to Ritta and Christina were side by side, and fused some time after closure of the vertebral column but before formation of the lower gut. In the case of the twins with fused faces the embryonic discs were head to head. The most extreme form of conjoined twinning is 'parapagus diprosopus', in which the fusion is so intimate that the only external evidence of twinning is a partly duplicated spinal column, an extra nose and, sometimes, a third eye. At this point all debates over individuality become moot.

Conjoined twins grade into parasites, infants that live at the expense of their siblings. The distinction is a matter of asymmetry. When the young Italian Lazarus Colloredo toured Europe in the 1630s he was celebrated for his charm and breeding even as his brother, John Baptista, dangled insensibly from his sternum. In the late 1800s an Indian boy, Laloo, displayed his parasite, a nameless, headless abdomen with arms, legs and genitals, in the United States. In 1982, a thirty-five-year-old Chinese man was reported with a parasitic head embedded in the right side of his own head. The extra head had a small brain, two weak eyes, two eyebrows, a nose, twelve teeth, a tongue and lots of hair. When the main head pursed its lips, stuck out its tongue or blinked its eyelids, so did the parasitic head; when the main head ate, the parasite drooled. Neurosurgeons removed it. Certain parts of the developing body seem especially vulnerable to parasitism, among them the neural tube, sternum and mouth. Some forty cases have been described of children who have dwarfed and deformed parasites growing from their palates. And parasites may themselves be parasitised. In 1860 a child was born in Durango, Mexico, who had a parasite growing from his mouth to which two others were attached.

Teratomas may be an even more intimate form of parasitism. These are disordered lumps of tissue that are usually mistaken initially for benign tumors, but that after surgery turn out to be compacted masses of differentiated tissue, hair, teeth, bone and skin. They have been traditionally blamed on errant germ cells. Unlike most of the body's cells, germ cells have the potential to become any other cell type, and it is supposed that occasionally a germ cell that has wandered into the abdomen will, perhaps by mutation, start developing spontaneously into a disordered simulacrum of a child. It is now suspected that some teratomas are, in fact, twins that have become fully enclosed within a larger sibling, a condition known trenchantly as '*foetus in foetu*'. A Dutch child born in 1995 had the remains of twenty-one foetuses (as determined by a leg count) embedded in its brain.

LEFT–RIGHT

There is one more thing that Ritta and Christina can tell us, and that is how we come to have a left and a right. We tend to think of ourselves as symmetrical creatures and, viewed externally, so

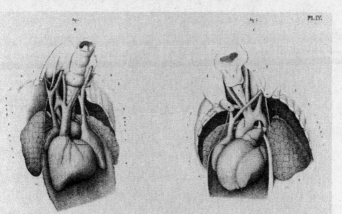

CONJOINED TWINS: SITUS INVERSUS VISCERA. RITTA AND
CHRISTINA PARODI. FROM ÉTIENNE SERRES 1832
Recherches d'anatomie transcendante et pathologique.

we are. To be sure, our right biceps may be more developed than
their cognates on the left (*vice versa* for the left-handed minority),
and none of us has perfectly matched limbs, eyes or ears, but
these are small deviations from an essential symmetry.
Internally, however, we are no more symmetrical than snails.
The pumping ventricles of our hearts protrude to the left sides
of our bodies. Also on the left are the arch of the aorta, the
thoracic duct, the stomach and the spleen, while the vena cava,
gall bladder and most of the liver are on the right. Christina's
viscera were arranged much as they are in any of us (except for
her liver, which was fused with Ritta's). Ritta's viscera, however,
were not. They were the mirror-image of her sister's.

This condition, known as situs inversus, literally 'position
inverted', is common in conjoined twins, as it is rare in the rest
of us (who are situs solitus). Not all conjoined twins are situs

inversus, but only those that are fused side to side (rather than head to head or hip to hip). Even among side-to-side twins situs inversus is only ever found in the right-side twin – 'right' referring to the twins themselves not the observer's view of them – and then only in 50 per cent of them. This last statistic is intriguing, for it implies that the orientation of the viscera is randomised in right-side twins. It is as if nature, when arranging their internal organs, abandons the determinism that rules the rest of us, and instead flips a coin marked 'left' or 'right'.

In recent years, much has been learned about why our internal organs are oriented the way they are. One source of information comes from those rare people – the best estimates put them at a frequency of 1 in 8500 – who, despite being born without a twin, have internal organs arranged the wrong way round. The most famous historical case of singleton situs inversus was an old soldier who died at Les Invalides in 1688. Obscure in life – just one of the thousands who, at the command of Louis XIV, had marched across Flanders, besieged Valenciennes and crossed the Rhine to chasten German princelings – he achieved fame in death when surgeons opened his chest and found his heart on the right. In the 1600s Parisians wrote doggerel about him; in the 1700s he featured in the *querelle des monstres* debate; in the 1800s he became an example of 'developmental arrest', the fashionable theory of the day. Were he to appear on an autopsy slab today, he would hardly be famous, but would simply be diagnosed as having a congenital disorder called 'Kartagener's syndrome'.

It is a diagnosis that allows us to reconstruct something of the old soldier's medical history. Although the immediate cause of his death is not known, it certainly had nothing to do with his inverted viscera. He was, indeed, in all likelihood oblivious to his own internal peculiarities. Although he was quite healthy (dying only at the age of seventy-two), he probably never fathered any children, and his sense of smell was also probably quite poor. We can guess these things because inverted viscera, sterility and a weak sense of smell are all features of men with Kartagener's.

That the association between these symptoms was ever noticed is surprising, for they seem so disparate, and even after the syndrome was first defined in 1936 the causal link between them remained elusive for years. But in 1976 a Swedish physician

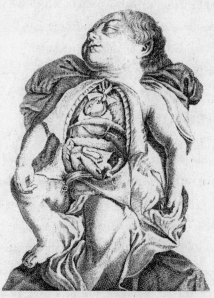

KARTAGENER'S SYNDROME. DISSECTED INFANT SHOWING SITUS INVERSUS VISCERA. FROM GEORGE LECLERC BUFFON 1777 *Histoire naturelle générale et particulière.*

named Bjorn Afzelius found that a poor sense of smell and sterility are caused by defective cilia – the minute devices that project from the surfaces of cells and wave about like tiny oars. Cilia clear particles from our bronchial passages, and the tail that drives a spermatozoon to its destination is also just a large sort of cilium. Each cilium is driven by a molecular motor, a motor that in people with Kartagener's syndrome does not work. As children, for want of beating cilia to clear the passages of their lungs and sinuses they have chronic bronchitis and sinusitis – hence the poor sense of smell. As adults, the men are sterile for want of mobile sperm. At the heart of the ciliary motor lies a large protein complex called dyenin. It is made up of a dozen-odd smaller proteins, each of which is encoded by its own gene. So far Kartagener's syndrome has been traced to mutations in at least two of these genes, and it is certain that others will be found.

But it is the situs inversus that is so intriguing. Afzelius noted that not all people with Kartagener's syndrome have inverted viscera: like conjoined twins, only half of them do. He suggested, insightfully, that this implied that cilia were a vital part of the devices that the embryo uses to tell left from right – but what their role was he could not say. Only in the last few years has the final link been made – and even now there is much that is obscure. It all has to do with (and this is no surprise) the organiser.

I said earlier that the organiser is a group of mesodermal cells located at one end of the embryo's primitive streak. Each of these cells has a single cilium that beats continually from right to left. Collectively they produce a feeble, but apparently all-important, current in the fluid surrounding the embryo, an amniotic Gulf Stream. This directional movement, and the cilia whose ceaseless activity causes it, is the first sign that left and right in the embryo are not the same. The mechanism, which was only discovered in 1998, is wonderfully simple and, as far as is known, is used nowhere else in the building of the embryo. What the cilia actually do is unclear; the best guess is that they concentrate some signalling molecule on the left side of the embryo, rather as foam accumulates in the eddies of a river.

This model (with its Aristotelian overtones) is frankly speculative, but it makes sense in the light of what happens next. Shortly after the organiser forms, genes can be seen switching on left and right in the cells that surround it. They encode signalling molecules that transmit and amplify the minute asymmetries established by the organiser's beating cilia to the rest of the embryo. One might call it a relay of signals, but that suggests something too consensual. It is more like a hotly contested election. In democracies left and right battle for the heart of the *polis*; so it is in embryos as well.

There is a lovely experiment that proves this. If the various signals that appear early in the embryo's life on either side of the organiser are indeed involved in helping it tell left from right, then it should be possible to confuse the embryo by switching the signals around. As usual, this is a hard trick to do in mammal embryos, but not that difficult in chickens. By gently cutting open a recently-laid egg and so exposing the embryo as it lies on its bed of yolk, it is possible to gently place a silicone bead soaked in 'left-hand' signal on its right (or to place a bead soaked in 'right-hand' signal on its left). Either way, the asymmetry of the embryo's signals is destroyed. And so too, it becomes apparent shortly thereafter, is the asymmetry of the chicken's heart. Where once it always fell to the left, it now has an even probability of falling to either side. The resemblance of this randomisation to that found in people with Kartagener's syndrome and in conjoined twins is surely no coincidence. Indeed, it is thought that Ritta's inverted heart was caused by just such a scrambled molecular signal. When the girls were nothing more than primitive streaks lying side by side, each strove to order her own geometry. But in Ritta's case this effort was confounded by signals that swept over from her left-hand twin. The molecular asymmetries upon which her future geometry depended were abolished, and from that point on the odds were fifty–fifty that her heart would be placed the wrong way round.

In 1974 Clara and Altagracia Rodriguez became the first conjoined twins to undergo successful surgical separation. Since then, the birth of each new pair – Mpho and Mphonyana (b.1988, South Africa), Katie and Eilish (b.1989, Ireland), Angela and Amy (b.1993, USA), Joseph and Luka (b.1997, South Africa), Maria Teresa and Maria de Jésus (b.2002, Guatemala), to name but a few – has been the occasion of a miniature drama in which surgeons, judges and parents have been called upon to play the part of Solomon. Surgical advances nothwithstanding, had Ritta and Christina Parodi been born today they could not have been separated. But they would surely have lived. Somewhere in America, Brittany and Abigail

Hensel, twins even more closely conjoined than they, have recently turned twelve.

Jules Janin never wrote his novel of Ritta and Christina Parodi's unlived lives. But he did leave an outline of what he had in mind. No translation could do justice to the turbulence of his prose, but a paraphrase gives an idea. In Janin's world, far from being born to poverty (after all, *'la misère gâte tout ce qu'elle touche'*), the two girls are rather well off. They also, inexplicably, have different-coloured hair. Christina, who is blonde, strong and noble, watches tenderly over her weaker, slightly sinister sibling, who is, inevitably, the brunette. All is harmonious, but suddenly seventeen springs have passed and, *arrive l'Amour*, in the shape of a bashful Werther who loves, and is loved by, only one of them – Christina, of course. Ah, the paradox! Two women, one heart, one lover; it is too tragic for words. Ritta sickens, and a mighty struggle between life and death ensues, as when *un guerrier est frappé à mort*. The sisters expire and we leave them having, as Janin puts it, 'arrived at new terrors, unknown emotions', and a sense of relief that he never wrote the full version.

The reality was, of course, quite different. When Serres had done with Ritta and Christina he not only kept the skeleton but quite a few other body-parts as well. An old catalogue of the Muséum d'Histoire Naturelle lists, in a copperplate hand, separate entries for the infants' brains (Cat. No. 1303 and 1304), eyes (1306, 1307), tongues (1308, 1309) and various other bits and pieces. Most of these specimens now seem to be lost, though it is possible that they will one day surface from the museum's underground vaults. Ritta and Christina's skeleton, however, is still around – as is the painted plaster-cast of their body. Both are on display in the Gallery of Comparative Anatomy, a steel-vaulted structure with an interior like a beaux-arts cathedral that stands only a few hundred metres from the amphitheatre where the sisters were first dissected.

A Gallery of Comparative Anatomy may seem like an odd place to exhibit the remains of two small girls. Nearly all of the hundreds of other skeletons there belong to animals, arranged by order, family, genus and species. Yet, from one point of view, there could hardly be a better place for them. The gallery represents the cumulative effort of France's greatest naturalists to impose order upon the natural world; to put each species where it should be; to make sense of them. For Étienne Geoffroy Saint-Hilaire the study of congenital deformity was, in the first instance, much in this spirit – a matter of locating conjoined twins in the order of things. In a gesture that Geoffroy would have loved, Ritta and Christina's remains share an exhibition case with a pair of piglets and pair of chicks that are conjoined much as they were. Such specimens were, to him, pickled proof that deformity is not arbitrary, a caprice of nature, a cosmic joke, but rather the consequence of natural forces that could be understood. 'There are no monsters,' he asserted, 'and nature is one.' In the way of French aphorisms, this is a little cryptic. But if you stand in front of the display case containing what is left of Ritta and Christina Parodi and look at the pink plaster-cast of the body with its two blonde heads and four blue eyes, it's easy to see what he meant.

III

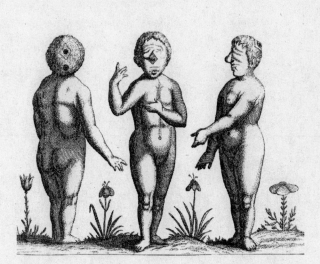

THE LAST JUDGEMENT

[ON FIRST PARTS]

IN 1890 THE CITIZENS OF AMSTERDAM bought Willem Vrolik's anatomical collection for the sum of twelve thousand guilders. It contained 5103 specimens, among them such rarities as the skull of a Sumatran prince named Depati-toetoep-hoera who had rebelled, apparently with little success, against his colonial masters. There was also a two-tusked Narwhal skull that had once belonged to the Danish royal family, an ethnographic collection of human crania, and the remains of 360 people displaying various congenital afflictions. Some of the specimens were adult skeletons, but most were infants preserved in alcohol or formaldehyde.

The Vrolik is just one of the great teratological collections that were built up during the eighteenth and nineteenth centuries. London's Guy's and St Thomas's Hospital has the

CYCLOPIA. STILLBORN INFANT, FIRME, ITALY (1624). FROM FORTUNIO LICETI 1634 *De monstrorum natura caussis et differentiis.*

Gordon collection, while the Royal College of Physicians and Surgeons has the Hunterian; Philadelphia has the Mütter; Paris the Muséum d'Histoire Naturelle as well as the Orfila and the Dupuytren. Vrolik's collection, which was given to the medical faculty of the University of Amsterdam, now occupies a sleek gallery in a modern biomedical complex on the outskirts of the city. What makes it unusual, if not unique, is that where most teratological collections are closed to all but doctors and scientists, the curators of the Vrolik have opened their collection to the public. In a fine display of Dutch rationalism they have decided that all who wish to do so should be allowed to see the worst for themselves.

And the worst is terrible indeed. Arrayed in cabinets, Vrolik's specimens are really quite horrifying. The gaping mouths, sightless eyes, opened skulls, split abdomens and fused or missing limbs seem to be the consequence of an uncontainable fury, as if some unseen Herod has perpetrated a latter-day slaughter of the innocents. Many of the infants that Vrolik collected were stillborn. A neonate's skeleton with a melon-like forehead is a case of thanatophoric dysplasia; another whose stunted limbs press against the walls of the jar in which he is kept has Blomstrand's chondrodysplasia. There is a cabinet containing children with acute failures in neural tube fusion. Their backs are cleaved open and their brains spill from their skulls. Across the gallery is a series of conjoined twins, one of which has a parasitic twin almost as large as himself protruding from the roof of his mouth. And next to them is a specimen labelled 'Acardia amorphus', a skin-covered sphere with nothing to hint at the child it almost became except for a small umbilical cord, a bit of intestine, and the rudiments of a vertebral column. Until one has walked around a collection such as the Vrolik's it is difficult to appreciate the limits of human form. The only visual referent that suggests itself are the demonic creatures that caper across the canvases of Hieronymus Bosch – another Dutchman – that now hang in the Prado. Of course, there is a difference in meaning. Where Bosch's grotesques serve to warn errant humanity of the fate that awaits it in the afterlife, Vrolik's are presented with clinical detachment, cleansed of moral value. And that, perhaps, suggests the best description of the Museum Vrolik. It is a *Last Judgement* for the scientific age.

THE CYCLOPS

Of Willem Vrolik's published writings, the greatest is a full folio work that he published between 1844 and 1849 called *Tabulae ad illustrandam embryogenesin hominis et mammalium tam naturalem quam abnormem* (Plates demonstrating normal and abnormal development in man and mammals). The teratological lithographs that it contains are of a beauty and veracity that have never been surpassed. The richest plates are those devoted to foetuses, human and animal, that have, instead of two eyes, only one – a single eye located in the middle of their foreheads. By the time Vrolik came to write the *Tabulae* he had been studying this condition for over ten years, had already published a major monograph on it, and had assembled a collection of twenty-four specimens – eight piglets, ten lambs, five humans and a kitten – that displayed this disorder in varying degrees of severity. Following Geoffroy he gave the condition a name that recalled one of the more terrible creatures in the Greek cosmology: the Cyclops.

Hesiod says that there were three Cyclopes – Brontes, Steropes and Arges – and that they were the offspring of Uranus and Gaia. They were gigantic, monstrous craftsmen who in some accounts made Zeus' thunderbolts, in others, the walls of Mycenae. The Cyclopes of the *Odyssey* are more human and more numerous than those of the *Theogony*, but their single eye is still a mark of savagery. Homer calls them 'lawless'. Polyphemus is more lawless than most: he has a taste for human flesh and dashes out the brains of Ulysses' companions 'as though they had been puppies' before eating them raw. Homer does not identify the island where the renegade Cyclops lived, but Ovid put him on the slopes of Etna in Sicily and gave him an affecting, if homicidal, passion for the nymph Galatea. Painted on vases, cast in bronze or carved in marble, Polyphemus was depicted by the Greeks throwing boulders or else reeling in agony as Ulysses drives a burning stick into his single eye.

Many teratologists have linked the deformity to the myth. They argue that the iconographic model for the semi-divine monster was a human infant. Certainly the model, if it ever existed, must have been only faintly remembered. Differences in size and vigour aside, even the earliest representations of

Polyphemus put his single eye where you would expect it, above his nose. But the single eye of a cyclopic infant invariably lies beneath its nose – or what is left of it. Others have argued, more or less plausibly, that the Cyclopes were inspired by the semi-fossilised remains of dwarf elephants that litter the Mediterranean islands.

Whatever its origins, Homer to Vrolik, the iconography of the Cyclops shows a clear evolutionary lineage. Homer's Polyphemus is monstrous; Ovid's is too, although he is also a sentimentalist. But within sixty years of the poet's death in 17 AD, the Cyclops would appear in a different guise. It would become a race of beings that had ontological status, supported by the authority of travellers and philosophers. In 77 AD Pliny the Elder finished his encyclopaedic *Historia naturalis*. Drawing on earlier Greek writers like Megasthenes, who around 303 BC travelled as an ambassador to India in the wake of Alexander the Great's conquests, Pliny peopled India and Ethiopia (the two were barely distinguished) with a host of fabulous races. There were the Sciapodes, who had a single enormous foot which they used as a sort of umbrella; dwarfish Pygmies; dog-headed Cynocephali; headless people with eyes between their shoulderblades; people with eight fingers and toes on each hand; people who lived for a thousand years; people with enormous ears; and people with tails. And then there were the single-eyed people: Pliny calls them the Arimaspeans and says that they fight with griffins over gold.

This was the beginning of a tradition of fabulous races that persisted for about fifteen centuries. By the third century AD,

CYCLOPS WOOING GALATEA. FROM BLAISE DE VIGENÈRE 1624
Les images Philostratus.

Christian writers had adopted the tradition; by the fifth century,
St Augustine is wondering whether these races are descended
from Adam. In the Middle Ages, the Cyclopes appears essentially unaltered from antiquity in manuscripts of wonder-books
such as Thomas à Cantimpré's *De Naturis Rerum* which was

composed around 1240. In the fourteenth century, their biblical parentage is settled: they are the deformed descendants of Cain and Ham. Around the same time they appear in illuminations of Marco Polo's travels (the Italian unaccountably fails to mention their existence); in the early 1500s one appears on the wall of a Danish church dressed in the striped pantaloons, floppy hat and leather purse of a late-medieval Baltic dandy. With time, the Cyclops becomes smaller, tamer and moves closer to home.

The first illustration of a cyclopic child, as distinct from a Cyclops, was given by Fortunio Liceti. In the 1634 edition of his *De monstrorum* he describes an infant girl who was born in Firme, Italy, in 1624 and who, he says, had a well-organised body but a head of horrible aspect. In the middle of her face, in place of a nose, there was a mass of skin that resembled a penis or a pear. Below this was a square-shaped piece of reddish skin on which one could see two very close-set eyes like the eyes of a chicken. Although the child died at birth she is depicted with the proportions of a robust ten-year-old, a legacy of the giants that preceded her.

Liceti describes another case of cyclopia as well, this time in a pair of conjoined twins whose crania are fused so that they face away from each other in true Janus style. Conjoined twinning and cyclopia is an unusual combination of anomalies, and one would be inclined to doubt its authenticity but for a 1916 clinical report of a pair of conjoined twins who showed much the same combination of features. And then there is the unusual provenance of Liceti's drawing. It is, he says, a copy of one preserved in the collection of His Eminence the Reverend Cardinal

Barberini at Rome, and the original, which now seems to be lost, was drawn by Leonardo da Vinci.

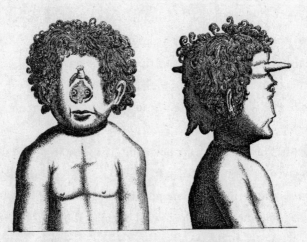

CYCLOPIA WITH CONJOINED TWINNING. ATTRIBUTED TO LEONARDO DA VINCI. FROM FORTUNIO LICETI 1634 *De monstrorum natura caussis et differentiis.*

Looking at his bottled babies, Willem Vrolik recognised that some were more severely afflicted than others. Some had only a single eyeball concealed within the eye-orbit, but in others two eyeballs were visible. Some had a recognisable nose, others had none at all. Modern clinicians recognise cyclopia as one extreme in a spectrum of head defects. At the other extreme are people whose only oddity is a single incisor placed symmetrically in their upper jaw instead of the usual two.

The single eye of a cyclopic child is the external sign of a disorder that reaches deep within its skull. All normal vertebrates have split brains. We, most obviously, have left and right cerebral hemispheres that we invoke when speaking of our left or right 'brains'. Cyclopic infants do not. Instead of two distinct cerebral hemispheres, two optic lobes and two olofactory lobes, their forebrains are fused into an apparently indivisible whole. Indeed, clinicians call this whole spectrum of birth defects the 'holoprosencephaly series', from the Greek: *holo* – whole, *prosencephalon* – forebrain. It is, in all its manifestations, the most common brain deformity in humans, afflicting 1 in 16,000 live-born children and 1 in 200 miscarried foetuses.

CYCLOPIA. STILLBORN CALF. FROM WILLEM VROLIK 1844–49
*Tabulae ad illustrandam embryogenesin hominis et
mammalium tam naturalem quam abnormem.*

The ease with which foetuses become cyclopic is frightening. Fish embryos will become cyclopic if they are heated, cooled, irradiated, deprived of oxygen, or exposed to ether, chloroform, acetone, phenol, butyric acid, lithium chloride, retinoic acid, alcohol or merely table salt. In the 1950s an epidemic of cyclopic lambs in the western United States was caused by pregnant ewes grazing on corn lilies, a plant of the subalpine meadows which has leaves rich in toxic alkaloids. In humans, diabetic mothers have a two-hundred-fold increased risk of giving birth to cyclopic children, as do alcoholic mothers.

Most cases of cyclopia are not, however, caused by anything the mother did (or did not do) during her pregnancy. Mutations in at least four and perhaps as many as twelve human genes also cause some form of holoprosencephaly. One of these genes encodes a signalling protein called sonic hedgehog. This molecule received its name in the early 1980s when a mutant fruit fly was discovered whose maggot progeny had a surplus of bristles covering their tiny bodies. 'Hedgehog' was the obvious name for

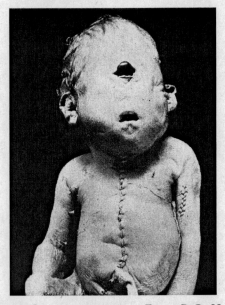

CYCLOPIA. STILLBORN INFANT. FROM B.C. HIRST AND G.A. PIERSOL 1893 *Human monstrosities.*

the gene, and when a related gene was discovered in vertebrates, 'sonic hedgehog' seemed the natural choice to a postgraduate student who perhaps loved his gaming-console too much. The sonic hedgehog mutations that cause cyclopia in humans are dominant. This implies that anyone who has just a single copy of the defective gene should have cyclopia or at least some kind of holoprosencephaly. But for reasons that are poorly under-stood, some carriers of mutant genes are hardly affected at all. They live, and pass the defective gene on to their children.

The fact that sonic hedgehog-defective infants have a single cerebral hemisphere tells us something important. When the forebrain first forms in the normal embryo it is a unitary thing, a simple bulge at the end of the neural tube – only later does it split into a left and right brain. This split is induced by sonic which, like so many signalling molecules, is a morphogen. During the formation of the neural tube, sonic appears in a small piece of mesoderm directly beneath the developing forebrain. Filtering up from one tissue to the next it cleaves the brain in two. This process is especially obvious in the making of eyes. Long before the embryo has eyes, a region of the forebrain is dedicated to their neural wiring. This region – the optic field – first appears as a single band traversing the embryo forebrain. Sonic moulds the optic field's topography, reducing it to two smaller fields on either side of the head. Mutations or chemicals that inhibit sonic prevent this – thus the single, monstrous, staring eye of the cyclopic infant.

But sonic does more than give us distinct cerebral hemispheres. Mice in which the sonic hedgehog gene has been completely disabled have malformed hearts, lungs, kidneys and guts. They are always stillborn and have no paws. Their faces are malformed beyond cyclopic, reduced to a strange kind of trunk: they have no eyes, ears or mouths. These malformations suggest that sonic is used throughout the developing embryo, almost anywhere it is growing a part. It even seems to be used repeatedly in the making of our heads.

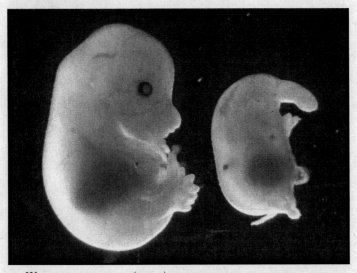

WILD TYPE MOUSE (LEFT); SONIC HEDGEHOG-DEFECTIVE
MOUSE (RIGHT).

An embryo's face is formed from five lumpy prominences that start out distinct, but later fuse with each other. Two of them become the upper jaw, two become the lower jaw, while one in front makes the nose, philtrum and forehead. These five prominences secrete sonic hedgehog protein. Sonic, in turn, controls their growth, and in doing so the geometry of the face. More exactly, it regulates its width. It sets the spaces between our ears, eyes and even our nostrils. We know this because chicken embryos whose faces are dosed with extra sonic protein develop unusually wide faces. If the dose is increased even further their faces become so wide that they start duplicating structures – and end up with two beaks side by side. Something like this also occurs naturally in humans. Several genetic disorders are marked by extremely wide-set eyes, a trait known as hyper-telorism. One of these is caused by mutations in a gene that nor-mally limits sonic's activity. Patients with another hypertelorism syndrome even resemble the sonic-dosed chickens in having very broad noses, or else noses with two tips, or even two noses.

Disorders of this sort prompt the question of just how wide a face can be. If, as a face becomes wider and wider, parts start duplicating, might one not ultimately end up with a completely duplicated face – and so two individuals? It is not an academic question. One Iowa-born pig arrived in the world with two snouts, two tongues, two oesophagi and three eyes each with an optic stalk of its own. It may have started out as two twin embryos that later conjoined in extraordinary intimacy. But given that the duplication was confined to the face and forebrain it may also have grown from a single primordial embryo, but one with a very wide head. The pig's head is preserved in a jar at the University of California San Francisco, a suitable object for philosophical reflection. Was it one pig or two? It's a question that would have stumped Aquinas himself. Not so the scientists who cared for the beast. They ignored the metaphysics, hedged their bets, and dubbed their friend(s) 'Ditto'.

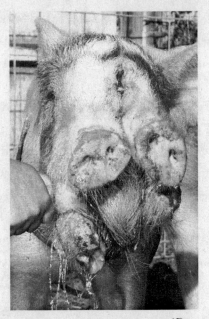

DUPLICATION OF FACE IN A PIG: 'DITTO'.

SIRENS

Among the disorders that appear regularly in the great teratology collections – the Vrolik devotes a whole cabinet to it – is a syndrome called sirenomelia. The name is taken from *siren*, the creatures that tempted Ulysses, and *melia*, for limb, but the English name, 'mermaid syndrome', is no less evocative. Instead of two good legs, sirenomelic infants have only one lower appendage – a tapering tube that contains a single femur, tibia and fibula. They resemble nothing so much as the fake mermaids concocted by nineteenth-century Japanese fishermen from the desiccated remains of monkeys and fish. More than Homeric echoes link cyclopia and sirenomelia. Just as cyclopia is a disorder of the midline of the face, a failure of its two sides to be sufficiently far apart, so sirenomelia is a failure in the midline of the lower limbs. A sirenomelic infant has neither a left nor a right leg but rather two legs that are somehow fused together.

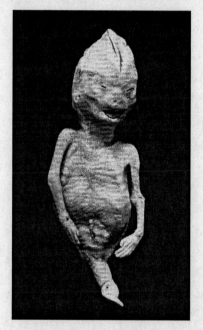

SIRENOMELIA OR MERMAID SYNDROME IN A STILLBORN FOETUS. FROM B.C. HIRST AND G.A. PIERSOL 1893 *Human monstrosities.*

The causes of sirenomelia are still not entirely known. But recently two groups of scientists independently engineered mouse strains that were defective for a particular gene. Unexpectedly, when the mice were born they had no tails and, just as sirenomelic infants do, fused hind limbs. To all appearances they were mermaid mice.

The mermaid mice were made by deleting the CYP26A1 gene. It encodes an enzyme that regulates a substance called retinoic acid. Most of the important molecules that control the construction of the embryo – that are a part of the genetic grammar – are proteins, long chains of amino acids. Retinoic acid, however, is not. Rather it is a much smaller and simpler sort of molecule, just a hydrocarbon ring with a tail. It is also one of the more mysterious of the embryo's molecules. Because it is not a protein it has been difficult to study. For one thing, it can't be seen in the embryo. The special stains that can be used to visualise proteins can't be used for hydrocarbon rings. And then, because it is not a protein there is no 'retinoic acid gene' – no single stretch of DNA that directly encodes the information needed to make it. Instead there are just genes which encode enzymes that manufacture retinoic acid or degrade it – a frustratingly indirect relationship between gene and substance.

Even so, there have long been hints that retinoic acid is important. Embryos manufacture their retinoic acid from vitamin A – the need of which has been clear since 1932, when a sow at a Texas agricultural college that had been fed a vitamin A-deficient diet gave birth to eleven piglets all of which lacked eyeballs. Conversely, the consequences of too much retinoic acid became apparent in the 1980s when a related molecule called isotretinoin was extensively prescribed for severe acne. The drug was taken orally, and though its teratogenic effects were by this time well known some women took it while unwittingly pregnant. In one study of thirty-six such pregnancies, twenty-three superficially normal infants were born, eight ended in miscarriages, and five infants were malformed, their defects including cleft palates, heart defects, disordered central nervous systems and missing ears.

Some scientists have tried to repeat this unplanned experiment by bathing animal embryos in retinoic acid and then looking for malformations. Often the outcome is just a miscellany of deformities, rather like those shown by isotretinoin-exposed infants. But sometimes the results can be spectacular. If a tadpole's tail is amputated, it normally grows another one in short order. But if the tail is amputated and the stump is painted with a solution of retinoic acid, the tadpole grows a bouquet of extra legs. This experiment clearly shows that retinoic acid is powerful stuff. It also suggests that tadpoles may use retinoic acid to regulate their rears. It does not, however, prove it. One could object that retinoic acid is, in effect, an exotic sort of poison, one that interferes in a completely unnatural way with the normal course of the embryo's progress.

Hence the importance of the mermaid mice. They give, for the first time, some real insight into what embryos use retinoic acid for. It seems it is a morphogen, one of the most important in the embryo. Indeed, one might almost call it an *Über*-morphogen that acts the length and breadth of the embryo. Being a hydrocarbon ring, however, it works rather differently from most other morphogens. Where protein-signalling molecules are too big to enter cells and so bind to receptors on their surfaces, retinoic acid penetrates the cell membrane and attaches to receptors within the cell that go right to the nucleus where they turn genes on and off.

Where does retinoic acid come from? And what, exactly, does it do? The CYP26A1 gene encodes an enzyme that degrades retinoic acid. Thus CYP26A1-defective mice have too much of it. Their mermaid-like limbs are caused by an anomalous surplus of retinoic acid in the embryo's rear. The rear of an embryo is not the only place affected by high levels of retinoic acid. Sirenomelic infants and mice also usually have head defects – implying that retinoic acid is normally lacking there too. Indeed, it is currently thought that could the concentration gradient of retinoic acid across an embryo be seen, it would resemble a hill with a peak somewhere near the embryo's future neck and slopes in all directions: sides, front and back. It would show a carefully constructed topography maintained by a balance of enzymes that make and degrade the morphogen, which in frogs with extra legs, mermaid mice, sirenomelic infants and foetuses exposed to acne-medications has been eroded away leaving only an ill-defined plateau.

THE CALCULATOR OF FATE

The morphogens that traverse the developing embryo – be they protein or hydrocarbon ring – provide cells with a kind of coordinate grid that they use to find out where they are and so what they should do and be. A cell is thus rather like a navigator who, traversing the wastes of the ocean, labours with sextant and chronometer to find his longitude and latitude. But there is one difference between navigator and cell: while the navigator's referents, the stars and planets, are always where they should be, the cell's sometimes are not. Sirenomelia and cyclopia are two instances where mutation has warped the universe that cells refer to or even caused its total collapse.

Yet even bearing this difference (inevitable when comparing the clockwork motions of the physical world with the jerry-built devices of biology) in mind, the analogy still has force. For all the constancy of the heavens, navigators have always lost their way – perhaps because the instruments by which they read the heavens become maladjusted. In the same way, the receptors which allow cells to perceive morphogens and measure their concentrations can also go awry – and any number of congenital disorders are caused by mutations that affect them.

But perhaps the deepest level of the analogy comes when we consider the calculations that navigators must make in order to establish where they are. Cells, too, calculate – and they do so with great precision, absorbing information from their environment, adding it up and arriving at a solution. This calculator – one might call it a calculator of fate – is composed of a vast number of proteins that combine their efforts within each cell to arrive at a solution. Of course, the calculator is not infallible: just as navigators occasionally get their sums wrong, so too, occasionally, do cells.

The consequences of cells making mistakes of this sort are beautifully illustrated by one of the more curious pieces of erotica dug from the ruins of Herculaneum. It is a small marble statue – no larger than a shoebox – that depicts Pan the goat-god, whom the Romans knew as Faunus, raping a nanny goat. Masterfully combining the animal and the human in equal parts, the unknown artist has given his Pan shaggy legs, cloven hooves, thick lips, a flattened snout and an expression of concentrated violence. He has also given the god an unusual anatomical feature. Suspended from his neck, just above the clavicles, are two small pendulous lobes that in life would be no more than a few centimetres long.

SUPERNUMERARY NECK AURICLES ON GOAT AND SATYR. *PAN RAPING A GOAT*. ROMAN COPY OF HELLENISTIC ORIGINAL, SECOND–THIRD CENTURY BC.

These lobes, which are very distinctive, only appear in Pans of the second or third century BC, or, as in this statue (now in the Secret Cabinet of the Naples Archaeological Museum), in later Roman copies of Greek originals. The innumerable goat-gods who chase across the black- or red-figure vases of the Classical period wooing shepherds or grasping at nymphs do not have them, nor do the allegorical Pans of the Renaissance and Baroque such as those in Sandro Botticelli's *Mars and Venus* or Annibale Carracci's *Omnia vincit Amor*. Neck lobes would also be quite out of place in the beautiful but vapid Pans of the Pre-Raphaelites.

The origin of the god's lobes is plain enough: they are echoed by an identical pair of appendages on his victim, the neck lobes frequently found on domesticated goats (German goatherds call them *Glocken* – bells). The sculptor of the original *Pan Raping a Goat* was clearly an acute observer of nature, and incorporated the lobes as one more detail to signify the goatishness of the god. Neck lobes, however, occur not only in goats but also, albeit rarely, in humans. In 1858 a British physician by the name of Birkett published a short paper describing a seven-year-old girl who had been brought to him with a pair protruding stiffly from either side of her neck. The girl had had them since birth. Birkett was not sure what they were, but he cut them off anyway and put them under the microscope, where he discovered that they were auricles – an extra pair of external ears.

SUPERNUMERARY AURICLES. EIGHT-YEAR-OLD GIRL, ENGLAND 1858. FROM WILLIAM BATESON 1894 *Materials for the study of variation.*

Extra auricles are an instance of a phenomenon called homeosis in which one part of a developing embryo becomes anomalously transformed into another. The particular transformation that causes neck-ears has its origins around five months after conception, when five cartilaginous arches form on either side of the embryo's head, positioned much where gills would be were the embryo a fish. Indeed, were the embryo a fish, gill arches are what they would become. In humans they form a miscellany of head parts including jaws, the tiny bones of the inner ears, and sundry throat cartilages. The visible, protuberant parts of our ears develop out of the cleft between the first and the second pair of arches. The remaining clefts usually just seal over, leaving our necks smooth, but occasionally in humans and often in goats, one of the lower clefts remains open and develops into something that looks much like an ear. The resemblance, however, is only superficial: the 'ears' have none of the internal apparatus that would enable them to hear.

Homeosis was first identified as a distinct phenomenon by the British biologist William Bateson, who in an 1894 book, *Materials for the study of variation*, coined the term and collected dozens of examples of such transformations. The *Materials* has something of the flavour of a medieval bestiary – Bateson called it his 'imaginary museum' – in which infants with supernumerary ears and heifers with odd numbers of teats jostle for space with five-winged moths, eight-legged beetles and lobsters that have antennae where their eyes should be. A strange book, then. Yet the *Materials* remains important to, and is cited by, molecular biologists in a way that few nineteenth-century zoological compendia are. This is because the transformations that Bateson identified pointed the way to one of the embryo's most beautiful devices: the genetic programme that permits cells, and so tissues and organs, to become different from each other. Homeosis pointed the way to the calculator of fate.

The calculator of fate was first discovered in fruit flies. Flies, like earthworms, are divided into repeating units or segments. These segments are especially obvious in maggots, though metamorphosis obscures some of their boundaries. Many segments in the adult fly are specialised in some way. Head segments carry labial palps (with which the fly feeds) and antennae (with which it smells); thoracic segments carry wings, legs, or small balancing organs called halteres; abdominal segments have no appendages at all. The organs of a given segment are established when the fly is only an embryo, long before they can actually be seen. To put it a bit more abstractly, in the embryo each segment is given an *identity*.

Over the last eighty-odd years, *Drosophila* geneticists have sought and found dozens of mutations that destroy the identities of segments. Some of these mutations cause flies to grow legs instead of antennae on their heads – and make a fly that cannot smell; others cause halteres to become wings – and make a four-winged dipteran that defies its own definition. Yet other mutations cause wings to become halteres – and leave the fly irredeemably earthbound.

These mutations disrupt a series of genes that, in homage to William Bateson, have come to be known as the homeotic genes.

There are eight of them, and they have names like Ultrabithorax, Antennapedia or, less euphemistically, 'deformed', that recall the strange flies produced when they are disrupted by mutation. They are the variables in a calculation that makes each segment distinct from any other.

The segmental calculator is a thing of beauty. It has the economical boolean logic of a computer programme. Each of the proteins encoded by the homeotic genes is present in certain segments. Some are present in the head, others in the thorax, others in the abdomen. The identity of a segment – the appendages it grows – depends on the precise combination of homeotic proteins present in its cells. The calculation for the third thoracic segment, which normally bears a haltere, looks something like this:

> IF *Ultrabithorax is* PRESENT
> AND *all other posterior homeotic proteins are* ABSENT
> THEN *third thoracic segment: HALTERE.*

Which simply implies that Ultrabithorax is necessary if the third thoracic segment is to grow a haltere, that is, to *be* a third thoracic segment. Should the gene be crippled by a mutation, the protein that it encodes, if present at all, will be unable to do its work. The segment's unique identity is lost; it becomes a second thoracic segment instead and carries wings.

When, in the 1980s, the homeotic genes were cloned and sequenced they proved to encode molecular switches: proteins that turn genes on and off. Molecular switches work by controlling the production of messenger RNA. Most genes contain information to make proteins. But this information requires a means of transmission. That is the job of messenger RNA, a molecule much like DNA except that it is neither double nor a helix, but only a long string of nucleotides. Messenger RNA is a copy of DNA, produced by a device that travels down gene sequences rather as a locomotive travels down a track. Molecular switches – or, to give them their proper name, 'transcription factors' – control this. Binding to 'regulatory elements', small, exact DNA sequences that surround every gene, transcription factors reach over to the molecular engine that makes messenger RNA and attempt to influence its workings. Some transcription factors seek to speed the engine up; others to shut it down. Attached to their regulatory elements, transcription factors face each other over the double helix and dispute for control. Like all negotiations, the outcome depends on the balance of power: the diversity of the opposing forces, or just their numbers.

The sequences of the eight fly homeotic genes are quite different. Yet each has a region, a sequence of only 180 base-pairs, that encodes, with small variations, the following string of amino acids:

RRRGRQTYTRYQTLELEKEFHTNHYLTRRRRIEM
AHALCLTERQIKIWFQNRRMKLKKEI.

This is the homeobox. In the sub-microscopic bulges and folds of a homeotic protein's three-dimensional topology it is the homeobox sequence, nestling within the grooves of the double helix of the DNA, that brings the homeotic proteins to their targets, the hundreds, perhaps thousands, of genes under their control. Subtle differences in the homeobox of each protein allows it to control particular suites of genes.

The discovery of the homeobox in 1984, distinctive as a Hapsburg's lip, suggested that the homeotic genes were all related to each other, that they were a family. Other animals, it quickly became apparent, had homeobox genes as well. They were found in worms and in snails, in starfish, fish, mice, and they were found in us. Perhaps they were present in the very first animals that crawled out of the Pre-Cambrian ooze a billion years ago. Most excitingly, if homeobox genes formed the circuits of the fly's calculator of parts, might they not do so for *all* creatures, even for humans? Molecular biologists are not a breed much given to hyperbole, but when they found the homeobox, they spoke of Holy Grails and of Rosetta Stones.

They were right to do so. Another of Vrolik's specimens, this time a skeleton, shows why. At first glance it seems a rather dull sort of skeleton. It isn't bent with rickets or bowed with achondroplasia; there is nothing unusual about it (though its skull, limbs and pelvis have evidently long gone astray). It is only an undulating vertebral column with brownish ribs on a rusted metal stand – an altogether abject thing. It is not even on display in the public galleries, but lives in a basement where it is shelved with dozens of other skeletons accumulated over a century but now largely surplus to requirements. And yet this skeleton enjoys a quiet renown. Each spring it sees the light of day as it is displayed to a new batch of the Rijkuniversiteit's medical students who are invited to identify its anomaly. This is surprisingly hard to spot, though obvious once pointed out – it is an extra pair of ribs.

Extra ribs have always caused trouble. In his *Pseudodoxia epidemica* Sir Thomas Browne relates how once, when the anatomist Renaldus Columbus dissected a woman at Pisa who happened to have thirteen ribs on one side, 'there arose a party that cried him down, and even unto oaths affirmed, this was the rib wherein a woman exceeded'. 'Were this true,' Browne continues, 'this would oracularly silence that dispute out of which side *Eve* was framed.' The influence of Genesis II: 21–22 on popular anatomy has been a baleful one. I recently asked a class of thirty biology undergraduates (among them Britain's best and brightest) whether men and women had the same number of ribs: about half a dozen of them thought not. 'But,' as Sir Thomas says with customary vigour, 'this will not consist with reason or inspection. For if we survey the Sceleton of both sexes, and therein the compage of bones, we shall readily discover that men and women have four and twenty ribs, that is, twelve on each side.' Just so. And yet extra ribs are surprisingly common: one in every ten or so adults has them (but they are no more or less frequent in women than men).

Most of us have thirty-three vertebrae. Starting at the head, there are seven neck vertebrae, then twelve rib-bearing vertebrae, then five vertebrae in the lower back, and another nine fused together to make the sacrum and coccyx or tail bones. In most people with extra ribs, this pattern is disrupted. A vertebra that normally does not bear ribs has become transformed into one that does. Sometimes this means the loss of a neck vertebra, sometimes the loss of one in the lower back; either way, homeotic transformations are much like the segment transformations that geneticists seek in their mutant flies.

It is no surprise, then, that the identity of each vertebra is controlled by homeotic genes much like those that keep a fly's segments in order. Of course, matters are rather more complicated for us. Flies have only eight homeotic genes while mammals have thirty-nine, so many that the evocative Latinate names have been dropped: no Ultrabithorax or proboscipedia for us, but only the prefix Hox followed by unmemorable letters and digits: Hoxa3, Hoxd13 and so on. In mammals, as in flies, homeotic genes begin their work early in the life of the embryo. Vertebrae develop from blocks of mesoderm called somites that form on either side of the nerve cord like rows of little bricks. Each homeotic protein is present in just some of the somites. All thirty-nine are present in the tail somites, but then they fall away, in ones and twos, so that finally only a handful remain in the somites closest to the head. The vertebral calculator is not very economical. For the seventh neck vertebra it looks something like this:

IF *Hoxa4 is* PRESENT

AND *Hoxa5 is* PRESENT

AND *Hoxb5 is* PRESENT

AND *Hoxa6 is* PRESENT

AND *Hoxb6 is* PRESENT

AND *all other posterior Hox genes are* ABSENT

THEN *a seventh neck vertebra will form: NO RIBS*

Should a mutation cripple any one of the genes that encode these five proteins, the seventh vertebra will transform into its neighbour, the eighth vertebra, and gain a pair of ribs.

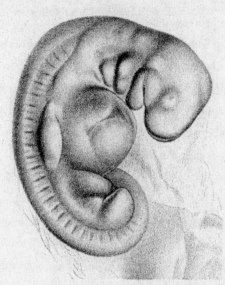

SOMITES IN A HUMAN EMBRYO. FROM FRANZ KEIBEL 1908
Normentafel zur Entwicklungsgeschichte des Menschen.

Distinguishing one vertebra from another is merely one instance of a problem that the embryo must solve repeatedly: the differentiation of parts along the head-to-tail axis. The embryo must solve this problem for the neural tube, uniform at first, but which later forms a brain at one end. It must solve it for the bones of the head – so that maxillae are formed next to mandibles and each is attached to its appropriate nerves and muscles. And it must solve this problem for the gut tube that becomes the stomach, liver, pancreas and intestines as well as the ventral blood vessel that becomes the four chambers of the heart. The Hox gene calculator is involved in all this.

How it works in mammals is known from mice in which one or more Hox genes have been deleted. Such mice are often profoundly disordered. Some have fore-limbs that are strangely close to their heads; others are missing parts of their hindbrains or cranial nerves. Some have hernias that cause their intestines to bulge into their thoracic cavities, or else open neural tubes. Some are missing their thymus, thyroid and parathyroid glands and have abnormal hearts and faces; some walk on their toes instead of on the soles of their feet, even as their hindquarters convulse uncontrollably. Most mice in which even one Hox gene has been deleted die young.

The Hox gene calculator is thought to work in humans in much the same way. The evidence for this belief is indirect and comes from a single 1997 study in which a group of London researchers stained six RU486 – 'morning after pill' – aborted embryos with molecular probes to reveal the times and place of homeotic gene expression. The embryos were four weeks old, about five millimetres long, and came from unwanted pregnancies. In autoradiographs of the sliced and stained embryos, Hox gene activity appears as grainy streaks and patches of white against the dark outlines of nascent rhombocephalons and pharyngeal arches. The patterns of Hox gene activity are just what one would expect from mice.

This is important and gratifying to know. But the study has not been repeated. Studies on human embryos are rare. In the United Kingdom they can only be done once formidable regulatory hurdles have been cleared; in the United States they can't be done at all, at least not in federally funded institutions. The autoradiographs that are the raw data of such studies certainly have a disquieting quality about them. Perhaps this is because in death these embryos reveal a property – gene activity – that truly belongs to the living.

THREE THOUSAND SWITCHES

Writing of the 'calculator of fate' I have emphasised the roles of the thirty-nine Hox genes. But the human genome encodes some three thousand other transcription factors. Like the signalling molecules to which they respond, transcription factors come in families, of which the homeobox genes are only one. These transcription factors are the circuit components, the switches if you will, that are thrown as cells calculate their fate. This computational process is a progressive one in which the earliest cells of the embryo, naive and confronted with a world of possibilities ahead of them, are ever more channelled into becoming one thing rather than another.

Some of these calculations, such as those that go into the vertebrae, are understood; others we are just learning about. In 1904, a Tyrolean innkeeper slaughtered one of the chickens wandering around his yard and found that it had no fewer than seven hearts. A curiosity? Perhaps. But in 2001 it was discovered that if a gene called β-catenin is deleted in mice, the result is an embryo with a string of extra hearts each of which beats and pumps blood. The extra hearts are made from tissue normally destined for the guts; and so a small part of another calculation – the one that decides whether a naive cell in the embryo becomes endoderm or mesoderm – stands revealed. Other disorders suggest the existence of calculations about which we know nothing. There is, it seems, a row of obscure glands in our eyelids (the Meibomian glands) that sometimes, albeit rarely, tranform into hair follicles. Infants who have lost their Meibomian glands have, instead, two or even three rows of eyelashes on each lid. It's a trait that runs in families, but the gene responsible for directing eyelid epidermis into a gland rather than a hair follicle has not yet been found (and one doubts that anyone is looking).

And then there is Disorganisation. A mouse mutant of unparalleled obscurity – it has been the subject of only three papers – it is also one of the strangest. Three properties make Disorganisation strange. The first is the pervasiveness of its effects upon the mice that carry it. It would be gratuitously macabre to detail the appearance of these mutant mice: it is enough to say that the deformities of a single litter would embrace the contents of a sizeable teratology museum. And yet, the mutation is not inevitably lethal. Disorganisation's second strange property is that no two mutant mice have the same set of defects. Some are hardly afflicted at all and can survive and breed, others are born mutilated but alive, yet others die in the womb. This variability extends to within a given mouse: a left kidney (or lung, or leg) may be destroyed even as its right cognate remains untouched. Finally, there is the strange propensity of the mutant mice to generate extra parts, not only supernumerary limbs (which can appear almost anywhere on the body), but also extra internal organs such as livers, spleens and intestines. They also have odd tumor-like structures embedded in their musculature and skin that seem to be the remains of supernumerary organs which never made it all the way. Is there a human Disorganisation gene? No human family showing Disorganisation-like properties seems to be known. However, some clinical geneticists have pointed to infants with especially bizarre suites of congenital anomalies as possible carriers of a cognate mutation. One such infant, a boy born in 1989, had nine toes on one foot and tumor-like pads of tissue scattered around his body. He also had a finger, complete with fingernail, growing from the right side of his ribcage. The Disorganisation gene has not yet been found, though it surely will be soon. Meanwhile, the mice speak. They tell of some critical, global, and quite unknown component of the embryo's calculator of fate, one that has gone utterly awry.

MUTATIS MUTANDIS

The power of the homeotic genes over the number and kinds of body parts has led some scientists to propose that they must be important in evolution; that they have somehow, worms to whales, provided animals with their staggering variety of forms. There may be something to this. People with extra ribs, specifically those who have extra ribs located on what should be their necks, are, for example, a bit like snakes. Snakes don't have necks at all: they have rib-bearing vertebrae that run all the way to their heads. This is because the pattern of Hox gene activity in the somites of snake embryos is quite different from that of necked reptiles, birds and mammals – a difference that also explains, incidentally, why snakes don't have arms. The position of arms, more generally fore-limbs, is dictated by the same Hox gene calculation that decides the allocation of vertebrae between neck and ribcage. No neck, no arms; it is as simple as that.

The beguiling quality of the homeotic genes has, however, less to do with differences among species than with similarities. These genes have a universality that is simply breathtaking. Flies use them to order their segments; we use them to sort out our vertebrae – but in both there is the common theme of ordering parts along the head-to-tail axis of the body. The similarities between the homeotic genes of vertebrates and insects also go far deeper than their general uses: they go right to the genome.

Homeotic genes come as clusters: groups of genes arrayed side by side on a single chromosome. The first few genes in the fly's homeotic cluster are involved in giving the head segments of the fly their identities; the next few genes along do the same for the thoracic segments; and the last few do the same for the abdominal segments. There is, it seems, a uncanny correspondence between the order of genes on the chromosome and the order of the fly itself. So, too, *mutatis mutandis*, is it for us. We have four clusters of homeotic genes on four chromosomes against the fly's one, but within each cluster the genes preserve the order along the chromosome that their cognates have in flies. Just as in flies, the first genes of each cluster are needed for our heads, the last for our tails, and the rest for the parts in between.

Why the homeotic genes should work in this way, and why they should have stayed doing so, is not clear. Nevertheless, they point to a system of building bodies that evolved perhaps as much as a thousand million years ago in some worm-like ancestor and that has been retained ever since. Indeed, the homeotic genes were merely the first indication that many of the molecular devices that make our bodies are ancient. Over the last ten years it has become plain that we are, in many ways, merely worms writ large. A gene called ems is needed to make a fruit fly's minute brain. So vast is the evolutionary gulf, both in time and complexity, between a fly's brain and the hundred-thousand-million-neuron edifice perched upon our own shoulders, that one could hardly expect that the same devices are used in both. Yet mutations in a human cognate of ems cause an inherited disorder that results in a brain abnormally riven with fissures (and so mental retardation and motor defects). Another fly gene called eyeless is needed to make a fly's compound eyes. Flies devoid of eyeless are, well, eyeless. So, in effect, are humans who inherit mutations in the cognate gene. They are born without irises.

In the cyclical way of intellectual fashion, all this has been said before, albeit far more obliquely. More than 150 years ago, that eccentric genius Étienne Geoffroy Saint-Hilaire – Linnaeus of deformity, discoverer of the universal law of mutual attraction – sought to construct a scientific programme, a *philosophie anatomique*, that would demonstrate that the animal world, seemingly so vast and various, was in fact one.

His initial goal was modest enough. Geoffroy attempted to show that structures that appear in mammals were the same, only modified, as those that appeared in other vertebrates, such as fish, reptiles and amphibians. In other words, he attempted to identify what we now call homologues, arguing, for example, that the opercular bones of fish (which cover the gills) were essentially the same as the tiny bones that make up the middle ears of mammals (the malleus, stapes and incus).

But opercular bones were small beer for a truly synthetic thinker: Geoffroy went on to find homologies between the most wonderfully disparate structures in the most wildly different creatures. Confronted with the exoskeleton of an insect and the vertebrae of a fish, he proposed that they were one and the same. To be sure, insects have an exoskeleton (all their guts inside their hard parts) while fish have an endoskeleton (bones surrounded by soft parts), but where other anatomists saw this as ample reason to keep them distinct, Geoffroy explained with the simple confidence of the visionary that 'every animal lives within or without its vertebral column'. Not content with this, he went on to show how the anatomy of the lobster was really very similar to that of a vertebrate – if only you flipped it on its back. Where lobsters carry their major nerve cord on their ventral sides (bellies) and their major blood vessels on their dorsal sides (backs), the reverse is true for vertebrates. And then there was the curious case of cephalopods: if one took a duck and folded it in half backwards so that its tail touched its head (an exercise performed, I believe, on paper alone), did its anatomy not resemble that of a cuttlefish?

It did not. Geoffroy's speculations attracted the wrath of Cuvier, his powerful rival at the Muséum. The result was a debate in front of the Académie Française in 1829 that Geoffroy lost — a duck doesn't look like a cuttlefish no matter how you bend it; even homologies between fish opercula and the mammalian middle ear didn't bear serious scrutiny. Yet if the particular homologies that he proposed sometimes seemed absurd, even in his day, his general method was not. Different organisms *do* have structures that are modified yet somehow similar. Indeed, the idea of homology is so commonplace in biology today (we speak of homology among genes as easily as among fore-limbs) that it is easy to read into Geoffroy's claims an evolutionary meaning he did not intend. The homologies that he saw, or thought he saw, were, as far as he was concerned, placed there by the Creator. It was the age of what would be called Transcendental Anatomy.

Today it is scarcely possible to study the development of any creature without comparing it to another. This is because animals, no matter how different they look, seem to share a common set of molecular devices that are the legacy of a common evolutionary history, that are used again and again, sometimes to different ends, but which remain recognisably the same wherever one looks. Indeed, the results of the genome sequencing projects suggest as much. Humans may have thirty thousand genes, but flies have thirteen thousand — a difference in number that is far smaller than one would expect given the seemingly enormous difference in size and complexity between the two species. Another creature much loved by developmental biologists, the nematode worm *Caenorhabditis elegans*, has nineteen thousand genes – even though the adult worms are only 1.2 millimetres long and have bodies composed of only 959 cells.

Some of Geoffroy's specific ideas are even being revived. One of these is his notion – on the face of it utterly absurd – that a vertebrate on its four feet is really just a lobster on its back. In the previous chapter I spoke of the signalling molecules that oppose each other to form the front and the back of vertebrate embryos. These same molecules – more precisely, their cognates or homologues – also distinguish back from belly in fruit flies; but with a twist. Where in a vertebrate embryo a BMP4 signal instructs cells to form belly, in flies the cognate molecule instructs cells to form back. And where in vertebrate embryos chordin instructs cells to form back, in flies the cognate molecule instructs cells to form belly. Somewhere in the evolutionary gulf that separates flies and mice there has, it seems, been an inversion in the very molecules that form the geometry of embryos, one that looks uncomfortably like the kind of twist that Geoffroy postulated. Absurd? Perhaps not. It is the sort of uncanny correspondence that one comes to expect in an age of Transcendental Genetics.

IV

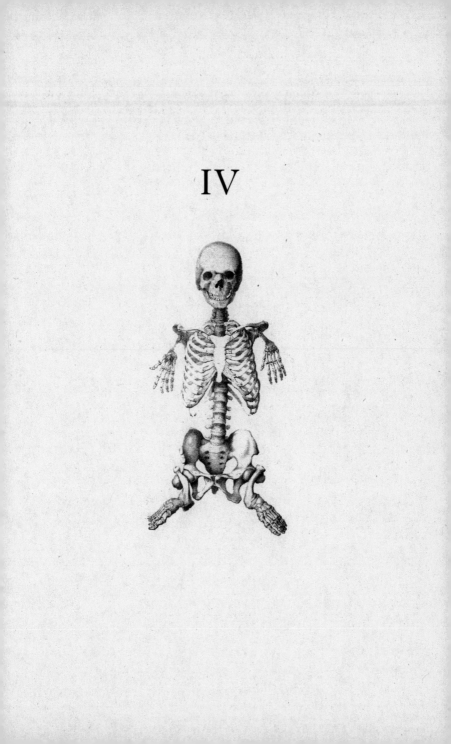

CLEPPIES

[ON ARMS AND LEGS]

OF ALL THE DOCTRINES THAT HAVE BEEN OCCASIONED by human deformity, none is more dismal than the belief that it is due to some moral failing. We can call this idea 'the fallacy of the mark of Cain'. For killing his brother, so Judeo-Christian tradition has it, God marked Cain and all his descendants. An apocryphal text from Armenia gives Cain a pair of horns; a Middle Irish history gives him lumps on his forehead, cheeks, hands and feet, while the author of *Beowulf* makes him the ancestor of the monstrous Grendel. None of this can actually be found in Genesis, which is, by comparison, a dull read. There Cain's punishment is exile, the mark is for his own protection, and its nature is left obscure. But then, the link between moral and

PHOCOMELIA. SKELETON OF MARC CAZOTTE, A.K.A. PEPIN (1757–1801). FROM WILLEM VROLIK 1844–49 *TABULAE AD ILLUSTRANDAM EMBRYOGENESIN HOMINIS ET MAMMALIUM TAM NATURALEM QUAM ABNORMEM.*

physical deformity has never really required biblical authority. It does not even require iniquitous parents. In 1999 the coach of the English national football team opined to an interviewer: 'You and I have been physically given two hands and two legs and a half-decent brain. Some people have not been born like that for a reason. The karma is working from another life. What you sow, you have to reap.' He took his cue from a Buddhist faith healer.

The fallacy of the mark of Cain flourished in Britain – football coaches aside – as recently as the seventeenth century. In 1685, in the remote and bleak Galloway village of Wigtown, two religious dissenters, Margaret McLaughlin and Margaret Wilson, were tried and convicted for crimes against the state. The infamy of their case comes from the cruelty of the method by which they were condemned to die. Both women were tied to stakes in the mouth of the River Bladnoch and left to the rising tide. Various accounts, none immediately contemporary, tell how they died. McLaughlin, an elderly widow, was the first to go; Wilson, who was eighteen years old, survived a little longer. A sheriff's officer, thinking that the widow's death-throes might concentrate the younger woman's mind, urged her to recant: 'Will you not say: God bless King Charlie and get this rope from off your neck?'

He underestimated the girl. Some accounts give her reply as a long and pious speech; others say she sang the 25th Psalm and recited Chapter 8 of Romans; all agree that her last words were pure defiance: 'God bless King Charlie, if He will.' The officer's response was to give vent to his talent for vernacular wit. 'Clep down among the partens and be drowned!' he cried. And then he grasped his halberd and drowned her.

The executioner's words are interesting. In the old Scottish dialect to 'clep' is to call; 'partens' are crabs. Thus: 'Call down among the crabs and be drowned.' In another version of the story, the officer was asked (by someone who had evidently missed the fun) how the women had behaved as the waters rose around them. 'Oo,' he replied in high humour, 'they just clepped roun' the stobs like partens, and prayed.' Either way, it is here that the story slides from martyrology into myth. For it seems that shortly after the officer – a man named Bell – had done his cruel work, his wife gave birth to a child who bore the ineradicable mark of its father's guilt: instead of fingers, its hands bore claws like those of a crab. 'The bairn is clepped!' cried the midwife. The mark of Bell's judicial crime would be visited on his descendants, many of whom would bear the deformity; they would be known as the 'Cleppie Bells'.

The spot at which the women are supposed to have died was marked by a stone monument in the form of a stake; today it stands in a reed-bed far from the water's edge, the Bladnoch having shifted course in the intervening three centuries. Another, far more imposing, monument to the martyrs stands on a hill above the town, and their graves, with carefully kept headstones, may be found in the local churchyard. Here, as elsewhere, the Scots nurse the wounds of history with relish.

There are other modern echoes of the event as well. As recently as 1900, a family bearing the names Bell or Agnew, and possessing hands moulded from birth into a claw-like deformity, lived in the south-east of Scotland and were said to be descendants of the Cleppie Bells. We know nothing more about them;

they may be there yet. We do know that in 1908 a large, unnamed family, living in London but of Scottish descent, were the subject of one of the first genetic studies of a human disorder of bodily form. Their deformity, known at the time as 'lobster-claw' syndrome, is certainly the same malformation that the Cleppie Bells had, though these days clinical geneticists eschew talk of 'lobster claws' and speak of 'split-hand-split-foot syndrome' or 'ectrodactyly', a term rendered palatable only by the obscurity of Greek, in which it reads as 'monstrous fingers'. This second Scots family may have been related to the Cleppie Bells, but it is quite possible that they were not and that the deformity arose independently in the two families. At one end of this story there is the historical trial and death of Margarets Wilson and

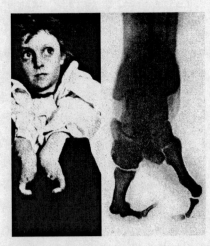

SPLIT-HAND-SPLIT-FOOT, OR ECTRODACYTLY, OR
LOBSTER-CLAW SYNDROME. GIRL WITH RADIOGRAPH
OF MOTHER'S FOOT, ENGLAND. FROM KARL PEARSON 1908
'ON THE INHERITANCE OF THE DEFORMITY KNOWN AS
SPLIT-FOOT OR LOBSTER CLAW'.

McLaughlin, at the other there are the Cleppie Bells and a clinical literature. The mythical element, of course, lies in the causal connection between the two. Nothing that officer Bell ever did could have caused his descendants to be born with only two digits on each hand, widely spaced apart. If the Bells were clepped, it was because some of them carried a dominant mutation that affected the growth of their limb-buds while they were still in the womb: it certainly had nothing to do with the partens.

THE USE OF LIMBS

The fragments of myth, folklore and tradition that remain to us from a pre-scientific age are like the marks left in sand by retreating waves: void of power and meaning, yet still possessed of some order. Muddied by time and confused causality, they still bear the imprint of the regularities of the natural world. It is surely significant that in such lore – no matter what its origin – few parts of the human body are as vulnerable to deformity as the limbs. Greek mythology has only one deformed Olympian, *crook-foot* Hephaestus who, abandoned by Hera (his mother), betrayed by Aphrodite (his wife), and spurned by Athena (his obsession), nevertheless taught humanity the mysteries of working metal and so is the god of craftsmen and smiths. Depicted on black-and-red-ware he is usually given congenital bilateral talipes equinovarus, or two club-feet. Oedipus, perhaps the most famous deformed mortal, wore his swollen foot in his name.

New myths arise even now. In the mid-1960s a Rhodesian Native Affairs administrator claimed that he knew of a tribe of two-toed people in the darker reaches of the Zambezi river valley. In tones reminiscent of Pliny the Elder's accounts of fabulous races in Aethiopia or the Indies they were, he said, variously called the Wadoma, Vadoma, Doma, Vanyai, Talunda or, most excitingly, the 'Ostrich-Footed People' – a primitive and reclusive group of hunter-gatherers who, by virtue of their odd feet, could run as swiftly as gazelles. Veracity was assured by a photograph of a Wadoma displaying his remarkable feet. In 1969 this same photograph appeared in the *Thunderbolt*, a newsletter published by the American National States Rights Party, illustrating an article which argued that since some Africans had 'animal feet' they were obviously a separate species ('Negro is related to Apes – Not White People'). American academics, rightly outraged, denounced the photograph as a forgery. Wrongly so, for when geneticists investigated the matter, they found that the Wadoma certainly existed, although far from being a whole tribe of 'ostrich-footed' people, there was only a single family afflicted with an apparently novel variety of ectrodactyly. But it is impossible to keep a good myth down. In the mid-1980s two South African journalists claimed they had stumbled across a whole tribe of two-toed people in the darker reaches of the Zambezi. Now, websites assert that the Wadoma worship a large metal sphere buried in the jungle and are, in fact, extraterrestrials.

Limbs have an extraordinary knack for going wrong. There are more named congenital disorders that affect our limbs than almost any other part of our bodies. Is it that limbs are particularly delicate, and so prone to register every insult that heredity or the environment imprints upon them? Or is it that they are especially complex? Delicate and complex they are, to be sure, but the more likely reason for the exuberant abundance of their imperfections is simply that they are not needed, at least not for life itself. Children may grow in the womb and be born with extra fingers, a missing tibia, or missing a limb entirely, and yet be otherwise quite healthy. They survive, and we see the damage.

One of the strange things about limbs is how easy it is to compensate for their absence, either partial or entire. As the patriarch of one ectrodactylous family replied to a geneticist: 'Bless 'e, sir, the kids don't mind it. They never had the use o' fingers and toes, and so they never misses 'em.' Indeed, why should they? They could hold their own at school in writing, drawing and even needlework. Among the adults, one was a bootmaker, one drove a cab, and another had a party trick in which he picked up pins from the floor using his two opposable toes.

The neural and physical versatility of limbs is even more striking in people who lack upper limbs altogether. Among the most engagingly feisty of all armless artists was Hermann Unthan, 'The Armless Fiddler'. Born in 1848 in a small German town, he narrowly escaped smothering by an infanticidal midwife, and was raised by his strict but loving parents on a diet of self-reliance that now seems positively heartless. Within days of his birth, his father ordered that his son was never to be pitied, never to be helped, and was not to be given any shoes or socks. By 1868 the young Hermann was giving violin recitals to delighted Viennese audiences as the younger Johann Strauss conducted. In the course of his long and varied life he travelled widely, finally coming to rest in the United States, which he loved. At the age of eighty he wrote his autobiography, aptly titled *The armless fiddler: a pediscript*, with his toes and an electric typewriter. This sort of neural flexibility is common in mammals. Among the anatomical wonders of the 1940s was a little Dutch goat that, born without fore-limbs, managed to get about bipedally, rather in the manner of a kangaroo.

THE NEAR TO THE FAR

The ability of animals to survive without their limbs has long proved useful to biologists. Limbs can be counted, dissected and manipulated on a living creature without the need to open the body. They are naked to the biologist's gaze. This visibility means that, of all the devices that make the body, those that make limbs are now exceptionally well understood. Much is known, for example, about their most salient characteristic: the fact that they stick out from our bodies.

At day 26 after conception, the first signs of a foetus's arms appear: two small bumps, one on each flank, just behind the neck. By analogy to the precursors of leaves or flowers, these bumps are called limb-buds. A day or so later, another pair of limb-buds forms further down the torso; they will become legs. Like any of the bumps on the surface of an embryo, limb-buds are at first just a bag of ectoderm filled with mesodermal cells. There are as yet no bones, muscles, tendons or blood vessels. The limb-bud remains in this amorphous state for about five weeks, at which time faint outlines of bones – the first signs of structure – begin to form. Even before that, however, the limb-bud has not been quiescent, because from nothing more than a small bump it has grown into an appendage about 2 millimetres long. On day 50 after conception, the embryo crouches and holds its newly formed hands over its heart. On day 56, it touches its nose.

What induces a limb-bud to grow out into space? In 1948 a young American biologist, John Saunders, gave an answer to this question. He had noticed that limb-buds were crowned by a ridge of unusual cells. The cells were clearly ectoderm – the tissue that covers the entire embryo – but at the tips of limbs they resembled tightly packed columns, quite unlike their usual pancake shape. Saunders dubbed this structure the 'apical ectodermal ridge' and then, curious to know more, decided to remove it.

As embryonic newts have been used to study the organiser, so chickens have been used to study limbs. Saunders operated on twenty-two foetal chickens, some young, others a little older. In each case he removed the apical ectodermal ridge from one wing-bud, while leaving the one of the other side intact. Having operated, he sealed up the egg and waited until the chicks hatched out. The operated wings all had a characteristic deformity: they were, to varying degrees, amputated. Chickens operated on when the limb-bud had just begun to expand showed severe amputations: they had at best a humerus (the bone closest to the shoulderblade), but below that, the radius, ulna, wrist bones and digits were all gone. Those operated on a little later had a humerus, radius and ulna, but lacked wrists and digits; later yet, only the digits were missing.

This experiment helps to explain why some infants, such as Hermann Unthan, are born without arms or legs. Our limb-buds also have apical ectodermal ridges, and sometimes they must surely fail. The ridges on Hermann's arm-buds probably malfunctioned soon after they first appeared; perhaps they never appeared at all. Other human deformities resemble the less extreme amputations seen in chicks whose wing ridges are removed only late in their growth. In the Brazilian states of Minas Gerais, São Paulo and Bahia there are families who are afflicted with a disorder called acheiropody – from the Greek:

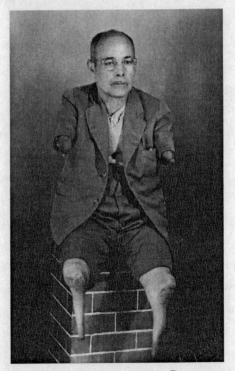

ACHEIROPODY. AN *ALEIJADINHO*, BRAZIL 1970S.

a – absence, *cheiros* – hand, *podos* – foot. Instead of hands and feet, the victims of this disorder have limbs that terminate in a tapered stump. They get about by walking on their knees and are called by the locals *aleijadinhos*, or 'little cripples'. The disorder is caused by a recessive mutation, probably quite an old one since it appears in more than twenty families, all of Portuguese descent. Because the mutation is recessive, only foetuses who have two copies of the mutant gene fail to develop hands and feet. Having two copies of a mutation is usually a sign of inbreeding: the first family of *aleijadinhos* ever studied were the children of a Peramá couple who were – local opinion varied – either full siblings, half siblings, or else uncle and niece.

The apical ectodermal ridge is the sculptor of the limb. As the development of the limb-bud draws to a close, the ridge regresses, leaving behind an outline of our fingers and toes. Should it be damaged in any way, the consequences will be visible in the limb's final form. The ectrodactylous hands of the Wigtown cleppies were the result of a mutation that caused a gap in the middle of the ridge, and so a gap in the middle of the forming limb. Mutations in at least four different genes are known to cause ectrodactyly, but it is quite possible that more will be discovered.

What gives the ridge, which is little more than a clump of cells, such power over the shape of a limb? The most obvious explanation would be that the cells making up the tissues of the limb – bone, sinew, blood vessels and so on – have their origin in the ridge. But this is not the case. All of these tissues are made of the mesoderm that lies beneath the ridge rather than the ridge itself; only the skin is ectoderm. The obvious alternative is that the ridge matters not as a source of cells, but rather as a source of information: it tells mesoderm what to do.

Action at a distance in the embryo usually implies the work of signals, and so it is in the limb-bud. Apical ectodermal ridges are rich in signalling molecules, especially so in one family of them: the fibroblast growth factors or FGFs. The experiment that identified FGFs as the source of the ridge's power began with the surgical extirpation, *à la* Saunders, of the apical ecto-dermal ridge from the tip of a young wing-bud. The denuded bud was not, however, allowed to grow up into the usual amputee wing. Instead, a silicone bead soaked in FGF was placed on its tip, more or less where the ridge would be. The result was a fully-grown limb – one cured, if you will, by the application of a single protein. Twenty-two genes in the human genome encode FGFs, of which at least four are switched on in the ridge. No one knows why so many are needed there, but col-lectively they are vital to the workings of the ridge. It would be an exaggeration to say that to grow a leg or an arm one needs only a little FGF, but clearly a little goes a long way.

Ridge FGFs not only keep mesodermal cells proliferating, they also keep them alive. Many cells will, at the slightest provo-cation, commit suicide. They have a whole molecular machinery to assist them in doing away with themselves. Seen through a microscope, a cell suicide is spectacular. Over the course of an hour or so the doomed cell becomes opaque, then suddenly shrivels and disappears as it is consumed by surrounding cells.

In the limb-bud, FGFs block the machinery of death; they give cells a reason to live. Yet while mass cell suicide is clearly a bad thing, at least some cell death is needed to form our fingers and toes, for if the ridge is the sculptor of the limb, cell death is the chisel. At day 37 after conception our extremities are as webbed as the feet of a duck. Over the next few days the cells in the webs die (as they do not in ducks) so that our digits may live free. Should a foetus have too much FGF signalling in its limbs, cells that should die don't. Such a foetus, or rather the child it becomes, has fingers and toes bound together so that the hand or foot looks as if it is wearing a mitten made of skin.

When Saunders removed the apical ectodermal ridge from a young limb-bud, the result was total amputation. Yet if the bud was older and larger, then only the structures further down – wrists, digits – were lost. Why? Over the last fifty years, various answers have been given to this question. The latest, though surely not the last, turns on two quite new observations. The first of these is that the ridge FGFs only penetrate a short way, about two hundred microns (one fifth of a millimetre) into the mesoderm. In a young limb-bud, two hundred microns-worth of suiciding cells cuts very deep as a proportion of total mass; in an older, larger limb-bud, much less so. This difference in proportion matters because limb-buds possess an invisible order. A limb-bud may look like an amorphous sack of cells, but even when newly formed, when it is no more than a bump on the foetal flank, its mesodermal cells have some foreknowledge of their fates. Some are already destined to become a humerus, others digits, yet others the parts between. As the limb-bud grows, each of these populations of cells proliferates and expands in turn. When a young limb-bud is deprived of FGFs, all of these variously fated cell populations suffer; when an older limb-bud is deprived only those closest to the tip do, and with them future hands and feet, toes and fingers.

This account of the making of our limbs contains within it the roots of twentieth-century medicine's most infamous blunder. In 1961 an Australian physician, William McBride, reported a sudden surge in the numbers of infants born with deformed limbs. Similar findings were reported a few months later by a German named Lenz. Both physicians suggested that the defects were caused by a sedative used to prevent morning sickness that has the chemical name phtalimido-glutarimide, but which swiftly became notorious by its trade-name, thalidomide. More reports rolled in from around the world. By the time it was all over, more than ten thousand infants in forty-six countries with thalidomide-induced teratologies had been found. Only the United States escaped the epidemic because a few sceptical FDA officials had delayed authorisation of a drug that was, at the time, the third best-selling in Europe.

The thalidomide infants had a very particular kind of limb deformity. Unlike acheiropods, their limbs did not suggest amputations in the womb, for most had reasonably formed hands and feet as well as shoulderblades and pelvises; they were simply missing everything else in between. Without long bones, their arms and feet connected almost directly to their torsos. Their limbs had the appearance of flippers – a condition dubbed phocomelia or 'seal-limb'.

Phocomelic infants have always appeared sporadically. In the sketchbooks of Goya (1746–1828), that compassionate connoisseur of deformity, there is a lovely sepia-wash portrait of a young mother proudly displaying her deformed child to two inquisitive old women. And there are, scattered throughout the early teratological literature, any number of people with the disorder. In his *Tabulae* (1844–49), Willem Vrolik gave a portrait of a phocomelic, a famous eighteenth-century Parisian juggler, Marc Cazotte, also known as 'Le Petit Pepin'. Vrolik also shows

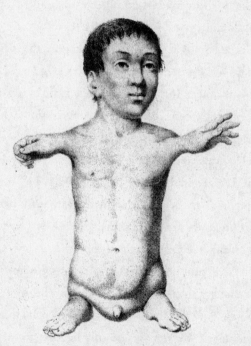

Phocomelia. Marc Cazotte, a.k.a. Pepin (1757–1801). From Willem Vrolik 1844–49 *tabulae ad illustrandam embryogenesin hominis et mammalium tam naturalem quam abnormem*.

Cazotte's skeleton, which still hangs in the Musée Duputryen in Paris, though its legs, by sad irony, are now missing. These cases of phocomelia might have been caused by some chemical or other, but they may also have been due to mutations, several of which cause the disorder. But until the 1960s, phocomelics were rare, little more than anatomical curiosities. Thalidomide turned them into icons of medical hubris.

How does thalidomide have its devastating effects? A comprehensive bibliography on the chemical and its consequences would run to about five thousand technical papers, but for all that, thalidomide is still poorly understood. Some things are clear. It is a teratogen and not a mutagen: the children of thalidomide victims are at no greater risk of congenital disorders than any others. Instead thalidomide inhibits cell proliferation. Taken by a pregnant woman during the time when she is most susceptible to morning sickness (thirty-nine to forty-two days after conception), it circulates throughout the bodies of mother and child and stops cells from dividing. This is when the earliest populations of cells that will form each part of the infant's future limbs are establishing themselves. Depending on the exact duration of the exposure, the precursors of one or more bones will fail to multiply; the result is a limb with missing parts. It is even thought that thalidomide may impede, quite directly, the fibroblast growth factors that are so essential to limb-bud development, but this remains speculation. Whatever its exact *modus operandi*, thalidomide is clearly a powerful drug and so a perennially attractive one. The taboo that surrounds it is breaking down as proposals for its use against a variety of diseases proliferate. In South America it is used to treat leprosy. Inevitably, infants with limb deformities are appearing once again as it is given to women who do not know that they have conceived.

GOING DIGITAL

Metric, with its base 10 units, exists only because the *savants* of the Académie Française who devised the system had ten fingers each on which they presumably learned to count. If pigs could do mathematics, they would probably measure their swill using a *Système International* devised from base 8, for they have only four digits per hoof. Horses have one digit per limb, camels have two, elephants have five, but guinea pigs have four on the forelimbs and three behind. Cats and dogs have five on the forefeet and five on the hind feet, but one of those is small, and is called a 'dew-claw'. Apart from some frogs and a kind of dolphin called a *vaquita*, most vertebrates never have more than five digits per limb.

Why this is so is deeply obscure. It is not as though extra digits are impossible to make. Mammals of all sorts sometimes show extra digits, but they are never common. St Bernards, Great Pyrenees, Newfoundlands and other large dogs are especially prone to having six digits on each foot – the duplication being an extra dew-claw. Ernest Hemingway's cats were polydactylous, and their many-toed descendants still live in the grounds of his Key West house. Fifteen per cent of the feral cats of Boston are polydactylous (some have up to ten extra toes), but there are no feral polydactylous cats in New York. There are many polydactylous strains of mice: one is called *Sasquatch* in homage to Big Foot, but most have more prosaic names such as *Doublefoot* or *Extra-Toes*. The American geneticist Sewall Wright once produced a baby guinea pig with forty-four fingers and toes in all, but it did not live.

And many people are born with extra digits. About 1 in 3000 Europeans is born with extra fingers or toes (or both), and about 1 in 300 Africans. Any digit can be duplicated, but in Africans it usually a little finger (pinkie), while in Europeans it tends to be a thumb. Polydactyly is usually genetic, frequently dominant, and can run for many generations in families. Long before Gregor Mendel ever lived, the French mathematician Pierre-Louis Moreau de Maupertuis (1698–1758) described the inheritance of polydactyly in the ancestors and descendants of a Berlin physician called Jacob Ruhe. Ruhe's grandmother had six fingers on each hand and six toes on each foot, as did his mother, as did he and three of his seven siblings, and two of his five children. Others have claimed even more impressive polydactylous pedigrees. In 1931 the Russian geneticist E.O. Manoiloff published an account of a polydactylous Georgian, Viačeslav Michailovič de Camio Scipion, who, he said, was able to document his descent from a lineage of polydactylous forebears reaching back six centuries.

If the apical ectodermal ridge ensures that our limbs grow out into space, another equally unobtrusive piece of limb-bud ensures that we have the right number and kinds of fingers. It was again John Saunders, along with a collaborator, Mary

Gasseling, who discovered it. They found that if they transplanted a piece of mesoderm from the tailmost edge of one chicken limb-bud onto the headmost edge of another (so that the bud had two tailmost edges in opposite orientation to each other), the result was a chicken wing with twice the usual number of digits. Most remarkably of all, the experimental wings were like a particularly exotic variety of polydactyly in humans. They resembled people who, far from having just an extra digit or two, have hands and feet that are almost completely duplicated with up to ten digits each. The polydactylous wings had a peculiar mirror-image geometry, one shared by duplicated hands in humans. If each finger is given a code in which the thumb is 1, forefinger 2, index-finger 3, ring-finger 4, and pinkie 5, then a normal, five-fingered, hand has the formula '12345', while a duplicated hand has the formula '5432112345'. It is that strangest of things, an anatomical palindrome.

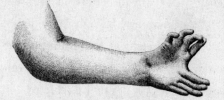

MIRROR-IMAGE POLYDACTYLY. LEFT HAND OF A WOMAN WITH EIGHT DIGITS. FROM WILLIAM BATESON 1894 *MATERIALS FOR THE STUDY OF VARIATION.*

Saunders and Gasseling called their potent piece of mesoderm the 'zone of polarising activity' or 'ZPA'. It is thought to be the source of a morphogen. At its source, where it is most concentrated, this morphogen induces naive mesoderm to become the little finger; further away, lower concentrations induce the ring, index, and forefinger in succession, and at the far opposite end of the limb, you get a thumb.

This account of how most of us come by our five fingers brings to mind the organiser. Like the organiser, the ZPA has the uncanny ability to impose order on its surroundings. And, just as the organiser morphogen was so eagerly sought for so long, so too, in recent years, has been the morphogen of the ZPA. It is almost certainly a signalling protein, likely a familiar one, a member of one of the great families of signalling proteins that also work elsewhere in the embryo. But limb-buds contain a plethora of such proteins, and it is hard to know which of them is the morphogen itself. In the past few years, several candidate molecules have been said to fit the bill. One of them is sonic hedgehog.

Sonic hedgehog appears in the limb-bud precisely where one would expect a morphogen to be: only in the mesoderm of the tailmost edge, exactly coincident with Saunders and Gasseling's ZPA. It also does what one would expect a morphogen to do: shape limbs. Chicken wings can be sculpted into new and improbable forms – including duplicate mirror-image poly- dactylous ones – simply by manipulating the presence of sonic in the bud. And then there are the mutants. Mutations in at least ten genes cause polydactyly in humans and all seem to affect, in some way or other, sonic's role in the limb.

But, as we saw in the previous chapter, sonic hedgehog does not just determine how many fingers and toes we have. It also divides our brains, decides how widely spaced our eyes will be, and regulates much else besides. It is an incorrigibly promiscuous molecule. Could we see the pattern of the sonic hedgehog gene's activity over time, as in time-lapse photography, we would see it flashing on and off throughout the developing embryo and foetus, now in this incipient organ, now in that one.

The devices responsible for all this have a formidable task, and nowhere, given sonic's power to direct the destiny of cells, do they have much room for error. These devices are transcription factors or 'molecular switches'. Some of them keep sonic in check. Should they be disabled by mutation, sonic turns on in parts of the limb-bud that it otherwise would not – and the result is extra fingers and toes. Other mutations do not disable the transcription factors themselves, but rather delete the regulatory elements to which they bind. The result, however, is the same: a confusion of morphogen gradients and an embarrassment of digits.

Polydactyly mutations relax control of sonic hedgehog, altering the balance of power in favour of ubiquity. But other mutations have exactly the opposite effect and prevent sonic from appearing in the limb-bud at all. The most blatant example of such a mutation is, of course, one that disables the sonic gene itself. Sonic-less mice have, in addition to their many other defects, no paws. This is strikingly reminiscent of a disorder that we have already come across: acheiropody, the disorder of the *aleijadinhos*. Indeed, there is some (disputed) evidence that the acheiropody mutation disables a regulatory element essential to sonic's presence in the limb.

This catalogue of mutations only hints at the complexities of gene regulation in the embryo. Whether or not a gene is turned on in a given cell depends on what transcription factors are found in that cell's nucleus, and their presence depends on the presence of yet other transcription factors, and so on. At first glance hierarchies of this sort seem to involve us in an infinite regression in which the burden of producing order is merely placed upon a previous set of entities which must, themselves, be ordered. But this dilemma is more apparent than real. The embryo's order is created iteratively. Sonic's precise presence in the ZPA is defined in part by the activity of Hox genes in the trunk mesoderm from which limbs grow. But the geometrical order that these genes give to the limb is crude; sonic's task is to refine it further. Beyond sonic there are, of course, yet further levels of refinement in which order is created on ever smaller scales, and each of them requires subtle and interminable negotiations, the nature of which we scarcely understand.

This vision of successive layers of negotiation and control may seem unimaginably complex. But in truth it is not complex enough, for it fails to capture one of the most pervasive properties of the embryo's programme: its non-linearities. I argued that the acheiropody mutation causes a failure of sonic to appear in the limb. And yet I began this chapter by arguing that infants with amputations in the womb, of whatever severity, were due to failures of the apical ectodermal ridge and the fibroblast growth factors they produce. This may seem like a contradiction, but it is only one if we think of the various limbs' signals as being independent of each other, when in fact they are not. For one of the most vital roles of sonic hedgehog is to maintain and shape the apical ectodermal ridge and its fibroblast growth factors; and one of the most vital roles of the apical ectodermal ridge is to maintain and shape the production of sonic hedgehog in the zone of polarising activity. There is a reciprocal flow of information as precarious as the flow of batons between two jugglers standing at opposite ends of a stage. Reciprocity of this sort is ubiquitous in the embryo and it alters the way we think about its growth and development. We begin with notions of linear pathways of command and control and simple geometries – and then watch as they unravel. For when, as in the limb, we actually begin to see the outlines of the embryo's programme, it invariably turns out to resemble a tangle of circuits that loop vertiginously across time and space. Circuits which, in this case, ensure that when we count our fingers and toes we usually come up with twenty.

HANDS, FEET AND ANCESTORS

Around day 32 after conception, when the human limb-bud is already well grown, its amorphous tissue begins to resolve into patterns. Ghostly precursors of bones appear: conglomerations of cells that have migrated together. The technical word for this process is 'condensation'. It hints at the way in which bones just quietly appear, rather like dew.

The first condensations to form become the bones closest to the body: the humerus in the arm, the femur in the leg. With time, conglomeration sweeps slowly down the limb-bud. The humerus divides into two new long, thin condensations, each of which will bud off by itself: the radius and the ulna. These condensations, in turn, divide and bud to form an arc of cells from which the twenty-seven bones of the wrist and palm are made. By day 38 after conception, the end of each limb-bud has become flat and broad, rather like a paddle. The paddle then folds into parallel valleys – four on each tip – leaving five islands of condensed cells: the future bones of the fingers and toes.

The shapes of the condensations depend, ultimately, on the reference grid laid down by the signalling systems of the limb. But, as elsewhere in the embryo, this information must be translated into cellular action. Hox genes do this for the head-to-tail axis of the embryo, and they also do it for the limb. As the limb-bud grows, some of the thirty-nine Hox genes appear in intricate overlapping patterns. They seem to be engaged in some combinatorial business analogous to the vertebral Hox code. Infants born with a single defective copy of the Hoxa13 gene have short big toes and bent little fingers. Another human Hox mutation causes synpolydactyly: extra fingers and toes fused together. A particularly devastating mutation that deletes no fewer than nine Hox genes in one go causes infants to be born with missing bones in the forearm, missing fingers and missing toes.

Limbs are not the only appendages in which Hox genes work. Infants born with Hox mutations that affect limbs tend to have malformed genitalia as well; in the worst cases male infants have just the vestiges of a scrotum and penis. Many of the molecules that make limbs also make genitals, and it should be no surprise that some mutations afflict both. The widely rumoured positive correlation between foot and penis size also, surprisingly, turns out to be at least partly true. No man should be judged by the size of his feet, however, for the correlation, though statistically significant, is weak. And then, such data as there are concern 'stretched' rather than erect penis length, surely the variable of interest. Still, when the French refer to the penis as *le troisième jambe*, *pied de roi* or *petit-doigt*; and the English to the *best-leg-of-three*, *down-leg* or *middle-leg*, not forgetting the optimistic *yard* which elsewhere means three feet, they speak truer than they know.

The Hox genes have also begun to tell us about origins. Where do fingers come from? It may seem that this question has a straightforward answer. Our limbs, flexible in so many dimensions, are the cognates of the structures that propel fish through the sea: their fins. But fish don't have fingers. One might suppose that the rays, those fine, bony projections that spread a fin like a fan, are their piscine equivalents. But fish rays and tetrapod digits are made of quite different kinds of bone – reason enough, anatomists say, to conclude that they have nothing to do with each other.

Most fish are only distantly related to tetrapods, so perhaps their want of fingers is no surprise. But even our closest piscine relatives are not much help. These are the lobe-finned fishes, among them the Australian lungfish, which spends much of its time buried in desiccated mud-flats, and the coelacanth, which inhabits the deeps of the Indian Ocean. Today's lobe-fins are often called 'living fossils', an allusion to the abundance of their relations four hundred million years ago and their scarcity now. Some fossil lobe-fins have fins that are strikingly like our own limbs; they seem to have cognates of a humerus, radius and ulna. They also have an abundance of smaller bones that look a bit like digits and that are made of the right kind of bone. But the geometry of these little bones is quite different to the stereotyped set of fingers and toes that is the birthright of all tetrapods. One can twist and turn a lung-fish's fin as much as one pleases, but the rudiments of our hands and feet simply do not appear. The conclusion seems unavoidable: fish don't have fingers, tetrapods do, and somewhere, around 370 million years ago, something new was made.

But how? Fish fin-buds are a lot like tetrapod limb-buds. They have apical ectodermal ridges, fibroblast growth factors, zones of polarising activity, sonic hedgehog, and panoplies of Hox genes that switch on and off in complicated ways as the bud pushes out into space. This tells us (what we already knew) that fins, legs and wings, so various in form and function, evolved from some *Ür*-appendage that stuck out from the side of some long-extinct *Ür*-fish.

We, however, are interested in the differences. One such difference lies in the details of the Hox genes. Early in the development of either a fin or a limb, Hoxd13 is switched on in the tailmost half, just around the zone of polarising activity. But as fins and limbs grow, differences begin to appear. In fish, the reign of Hoxd13 is brief; as the fin-buds grow it just gradually fades away. In mice, however, Hoxd13 stays on in an arc that stretches right across the outermost part of the limb. It seems to be doing something new, something that is not, and never has been, done in fish: Hoxd13 is specifying digits.

Such differences (which are true of other Hox genes as well) give Hox gene mutations their deeper meaning. If, in its last flourish of activity, Hoxd13 is specifying digits, one would expect that a mouse in which Hoxd13 has been deleted would be a mouse with no digits. It would be a mouse in which just one of the many layers of change that have accreted over the course of five-hundred-odd million years of evolution has been stripped away. Its paws would be atavistic: incrementally less tetrapod-like and incrementally more fish-like. As it turns out, however, Hoxd13-mutant mice, far from having a lack of digits, have a surplus of them. Their digits are small and crippled, but instead of the usual five, they also have a sixth.

This result is rather puzzling. It seems to suggest that something, somewhere, in our evolutionary history not only had fingers and toes, but had more of them than we, and nearly all living tetrapods, do. The idea that polydactyly (be it in mice, guinea pigs, dogs, cats or humans) is an atavism is an old one. Darwin claimed as much in the first edition of his *The variation of animals and plants under domestication* (1868), a work in which he attempted to develop the theory of inheritance that evolution by natural selection so badly needed. 'When the child resembles either grandparent more closely than its immediate parent,' he wrote, 'our attention is not much arrested, though in truth the fact is highly remarkable; but when the child resembles some remote ancestor or some distant member of a collateral line, – and in the last case we must attribute this to the descent of all members from a common progenitor, – we feel a just degree of astonishment.'

This is certainly true, but Darwin's reasons for thinking that polydactyly in humans is an atavism (or 'reversion' to use his terminology) are, to say the least, obscure. Salamanders, he noted, could regrow digits following amputation, and he had read somewhere that supernumerary fingers in humans could do the same thing even if normal ones could not. Extra digits were somehow, then, the product of a primitive regenerative ability, and hence atavisms.

It was a woolly argument, and it did not go unchallenged. The German anatomist Carl Gegenbauer pointed out that human fingers, supernumerary or otherwise, could not regenerate if amputated, and even if they could, so what? Polydactyly could not be an atavism without a polydactylous ancestor, and all known tetrapods, living or dead, had five fingers. In the next edition of *The variation* seven years later, Darwin, ever reasonable, admitted that he'd been wrong: polydactylous fingers weren't atavisms; they were just monstrous.

But Darwin may have been right after all – albeit for the wrong reasons. In the last ten years or so, the ancestry of the tetrapods has undergone a radical revision. New fossils have come out of the rocks, and strange things are being seen. Contrary to all expectations, humans – and all living tetrapods – *do* have polydactylous ancestors. The earliest unambiguous tetrapods in the fossil record are a trio of Devonian swamp-beasts that lived about 360 million years ago: *Acanthostega*, *Turlepreton* and *Ichthyostega*. All of them are, by modern tetrapod standards, weirdly polydactylous: *Acanthostega* has eight digits on each paw, *Turlepreton* and *Ichthyostega* have either six or seven. Suddenly it seems quite possible that Hoxd13-mutant mice, and mutant polydactylous mammals of all sorts, are indeed remembrances of times past – only the memory is of an early amphibian and not a fish.

Perhaps more genetic fiddling is required to get back to a fish fin; more layers have to be removed. This seems to be so. Mice that are mutant for Hoxd13 may be polydactylous, but mice that are mutant for Hoxd13 as well as other Hox genes – that is, are doubly or even trebly mutant – have no digits at all. It may be that as developmental geneticists strip successive Hox genes from the genomes of their mice, they are reversing history in the laboratory; they are plumbing a five-hundred-million-year odyssey that reaches from fish with no fingers to Devonian amphibians with a surplus of them, and that ends, finally, with our familiar five.

V

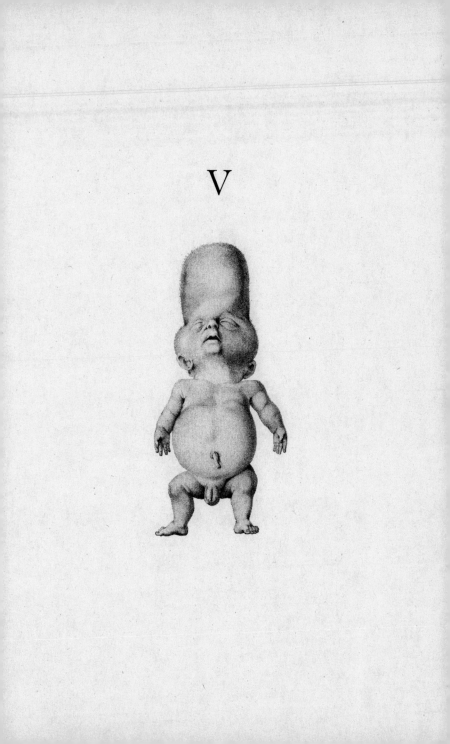

FLESH OF MY FLESH,
BONE OF MY BONE

[ON SKELETONS]

AROUND 1896, a Chinese sailor named Arnold arrived at the Cape of Good Hope. We do not know much about him, nor are there any extant portraits. We can, however, suppose that he was rather short and that he had a bulging forehead. He was probably soft-headed – not a reflection on his intelligence, but rather on the fact that he was missing the top of his skull. He probably did not have clavicles, or if he did, they may not have made contact with his shoulderblades. Had someone stood behind him and pushed, Arnold's shoulders could have been induced to meet over his chest. He may have had supernumerary teeth or he may have had no teeth at all.

THANATOPHORIC DYSPLASIA. STILLBORN INFANT, AMSTERDAM C.1847. FROM WILLEM VROLIK 1844–49 *TABULAE AD ILLUSTRANDAM EMBRYOGENESIN HOMINIS ET MAMMALIUM TAM NATURALEM QUAM ABNORMEM.*

We can guess all this because Arnold was exceptionally philoprogenitive, and many of his numerous descendants carry these traits. Arriving in Cape Town, he converted to Islam, took seven wives, and submerged himself in Cape Malay society. The Cape Malays are a community of broadly Javanese descent, but one that has absorbed contributions from San, Xhoi-Xhois, West Africans and Malagasys within its genetic mix. Traditionally artisans and fishermen, the Cape Malays made the elegant gables of the Cape Dutch manors found on South Africa's wine-growing estates, gave the nation's cuisine its Oriental tang, and the Afrikaans language a smattering of Malay words such as *piesang*. A 1953 survey revealed Arnold's missing-bone mutation in 253 of his descendants. By 1996, the mutation had been transmitted to about a thousand people. Fortunately, a lack of clavicles and the occasional soft skull are not very disabling. Arnold's clan are, indeed, quite proud of their ancestor and his mutation.

MAKING BONE

Perhaps because they are the last of our remains to dissipate to dust, we think of bones as inanimate things. But they are not. Like hearts and livers, bones are continually built up and broken down in a cycle of construction and destruction. And though they seem so separate from the rest of our bodies, they originate from the same embryonic tissues that make the flesh that covers them. In a very real sense, bone is flesh transformed.

The intimate relationship between bones and flesh can be seen in the origin of the cells that make them. Most bone cells –

osteoblasts – are derived from mesoderm, the same embryonic tissue that also gives rise to connective tissue and muscle. The relationship can also be seen in the way that bones form. Buried within each bone are the remains of the cells that made it.

Our various bones are made in two quite different ways. Flat bones, such as those of the cranium, start out in the embryo as a layer of osteoblasts that secrete a protein matrix. Calcium phosphate spicules form upon this matrix and encase the cells. As the bone grows, layers of osteoblasts are added and each is, in turn, entombed by its own secretions. Long bones, such as femurs, do things a bit differently. They start out as the condensations of cells that are visible in an embryo's developing limbs. These cells, which are also derived from mesoderm, are called chondrocytes and they produce cartilage. The cartilage is a template for the future bone, one that only later becomes invaded by osteoblasts. When the template first appears, it is bone in form but not in substance.

One of the molecules that controls these condensations is bone morphogenetic protein (BMP). It is convenient to speak of it as one molecule, but it is really a family of them. Like so many families of signalling molecules, the BMPs crop up in the most unexpected places in the embryo. It is a BMP that, long before the bones are formed, instructs some the embryo's cells to become belly rather than back. In older embryos, however, BMPs appear in the condensations of cells that will become future bones. In children and adults, they appear around fractured bones. The remarkable thing about BMPs is their ability to induce bone almost anywhere. If one injects BMPs underneath the skin of a rat, nodules of bone will form that are quite detached from the skeleton, but that look very much like normal bone, even to the extent of having marrow.

To make bone it is not enough that undifferentiated cells condense in the right places and quantities. The cells have to be turned into osteoblasts and chondrocytes. To return to a metaphor that I used earlier, they have to calculate their fates. The gene that calculates the fates of osteoblast happens to be the one responsible for 'Arnold-head'. This gene encodes a transcription factor called CBFA1. It may be thought that CBFA1 is not very important, since mutations in it result only in a few missing bones. However, Arnold's descendants are heterozygous for the mutation: only one of their two CBFA1 genes carries the mutant copy. Mice heterozygous for a mutation in the same gene also have soft heads and lack clavicles. But mice that are homozygous for the mutation are literally boneless. Instead of skeletons they have only bands of cartilage threading through their bodies, and their brains are protected by little more than skin. They are completely flexible and they are also dead. Boneless mice die within minutes of being born, asphyxiated for want of a ribcage to support their lungs.

By one of those quirks of genetic history, South Africa is also home to a mutation that has the opposite effect of Arnold's: one that causes not a deficiency of bone, but rather an excess. Far from having holes in their skulls, the victims of this second mutation have crania that are unusually massive. The mutation's effects are not obvious at birth. The thick skulls and coarse features that characterise this syndrome only come with age. Unlike the boneless mutation, the extra-bone mutation is often lethal. Its victims usually die in middle age from seizures as the excess bone crushes some vital nerve. Again, unlike the boneless mutation, the thick-skull mutation is recessive and so is expressed in only a handful of people – inbred villagers descended from the original Dutchmen who founded the Cape Colony in the seventeenth century.

The mutation that causes this disorder disables a quite different sort of gene from CBFA1. The protein itself is called sclerostin, after the syndrome sclerosteosis. It is thought to be an inhibitor of BMPs – perhaps it binds to them and so disables them. This is how many BMP inhibitors work. In the early embryo, organiser molecules such as noggin restrict the action of BMP in just this way. Indeed, noggin mutations are responsible for yet another bone-overgrowth syndrome that affects only finger-bones and causes them to fuse together with age, rendering them immobile.

Surplus-bone disorders illustrate the need that our bodies have to keep BMPs under control. Yet fused fingers and even thick skulls are relatively mild manifestations of the ability of BMPs to produce bone in inconvenient places. Another disease shows the extent of what can go wrong when osteoblasts proliferate throughout the body and make bone wherever they please. The disorder is known as fibrodysplasia ossificans progressiva or FOP. It is rare: estimates put the number of people afflicted with it worldwide at about 2500, but only a few hundred are actually known to specialists in the disease. Its most famous victim was an American man by the name of Harry Raymond Eastlack. In 1935, Harry, then a five-year-old, broke his leg while playing with his sister. The fracture set badly and left him with a bowed left femur. Shortly afterwards, he also developed a stiff hip and knee. The stiffness was not, however, caused by the original break, but rather by bony deposits that had grown on his adductor and quadriceps muscles.

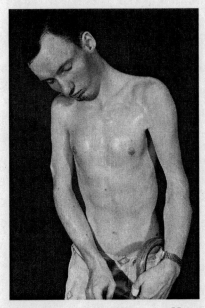

FIBRODYSPLASIA OSSIFICANS PROGRESSIVA. HARRY EASTLACK, USA 1953.

As Harry grew older, the bony deposits spread throughout his body. They appeared in his buttocks, chest and neck and also his back. By 1946 his left leg and hip had completely seized up; his torso had become permanently bent at a thirty-degree angle; bony bridges had formed between his vertebrae, and the muscles of his back had turned to sheets of bone. Attempts were made to surgically excise the bone, but it grew back – harder and more pervasive than before. At the age of twenty-three, he was placed in an institution for the chronically disabled. By the time of his death in 1973, his jaws had seized up and he could no longer speak.

Harry Eastlack requested that his skeleton be kept for scientific study, and today it stands in Philadelphia's Mütter Museum. Bound in extra sheets, struts and pinnacles of bone that ramify across the limbs and ribcage, the skeleton is, in effect, that of a forty-year-old man encased in another skeleton, but one that is inchoate and out of control. The cause of the disease is

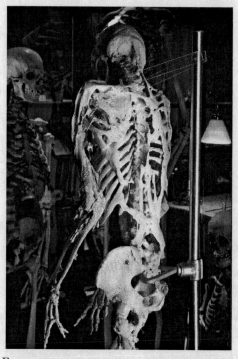

FIBRODYSPLASIA OSSIFICANS PROGRESSIVA.
HARRY EASTLACK (1930–73).

understood in general terms. The bodies of FOP patients do not respond to tissue trauma in the normal way. Bruises and sprains, instead of being repaired with the appropriate tissue, are repaired with osteoblasts and the new tissue turns to bone. This has all the hallmarks of an error in BMP production or control, but the mutation itself has not yet been identified. The search may well be a long one. FOP patients rarely have children, so the causal gene cannot be mapped by searching through long pedigrees of afflicted families.

GROWING BONES

A newly born infant has a skeleton of filigree fineness and intricacy, a skull as soft as a sheet of cardboard but scarcely as thick, and femurs as thin as pencils. By the time the child is an adult all this will have changed. The femur will have the diameter of a hockey stick, and will be able to resist the impact of one as well, at least most of the time. The skull will be as thick as a soup plate and capable of protecting the brain even when its owner is engaged in a game of rugby or the scarcely less curious customs of the Australian Aborigines who ritually beat each other's skulls with thick branches.

What makes bones grow to the size that they do? In 1930 a young American scientist, Victor Chandler Twitty, tackled this question in a very direct way. Taking a cue from the German *Entwicklungsmechanik*, Twitty chose to study two species of salamanders: tiger salamanders and spotted salamanders. Closely related, they differ in one notable respect: tiger salamanders are about twice as big as spotteds. The experiment he carried out on them was of such elegance, simplicity and daring that seventy years later it can still be found in textbooks.

Twitty began by cutting the legs off his salamanders. The Italian scientist Lazzaro Spallanzani of Scandiano had discovered in 1768 that salamanders can regrow, should they need to, their legs and tails. Since then, thousands of the creatures have lost their legs to science. One luckless animal had a leg amputated twenty times – and grew it back each time. It is sometimes facetiously remarked among scientists that happiness is finding an experiment that works and doing it over and over again. Twitty, however, was more ingenious. As the stumps of his salamanders healed, and as their tissues reorganised into limb-buds, he once again put them to the knife. He then took the severed limb-buds of each species and grafted them onto the stumps of the other.

The question was, how big would the foreign limbs grow? There were, Twitty reasoned, two possibilities. As the grafted buds grew into legs, they might take on the properties of their host, or they might retain their own. If the first, then a spotted salamander limb-bud grafted onto a tiger salamander should grow into a hefty, tiger salamander-sized leg. Alternatively, the spotted salamander limb-bud might simply grow into the small leg that it usually does. The result would be tiger salamanders with three large legs and one tiny grafted one, and spotted salamanders with three tiny legs and one large grafted one – in short, lopsided salamanders.

Twitty expected that the foreign legs would grow as large as the host salamanders' normal legs. By the 1930s it was known that hormones have an immense influence over human growth. One, produced by the pituitary gland, had even been dubbed 'growth hormone', and clinicians spoke of people with an excess or deficiency of this hormone as 'pituitary' giants and dwarfs. If tiger salamanders were larger than spotted salamanders, it was surely because they had more growth hormone (or something like it) than their smaller relatives. Foreign limbs should respond to the hormone levels of their hosts no less than ordinary limbs and should become accordingly large or small. The control of growth would be, in a sense, global – a matter of tissues being dictated to by a single set of instructions that circulate throughout the whole body.

There is no doubt that hormones do play a role – a vital role – in how large salamanders, people, and probably all animals become. But the beauty of Twitty's experiment is that it showed that, however important hormones are, they are not responsible for the difference between large and small salamanders. Against expectation, his salamanders proved lopsided. It seemed as if the grafted limbs, in some ineffably mysterious way, simply knew what size they should be regardless of what they were attached to. It was an experiment that showed the primacy of the local over the global, and that each salamander leg contains within itself the makings of its own fate.

The reward of these experiments was, for Twitty, enduring fame of a modest sort. More immediately, in 1931 he got to go to Berlin. He went to work at the laboratory of Otto Mangold, husband of Hilda Pröscholdt of organiser fame, at the Kaiser Wilhelm Institute. There he met some of the great biologists of the day: Hans Spemann, Richard Goldschmidt and Viktor Hamburger, who together had made Germany pre-eminent in developmental biology. Neither Twitty's research at the Kaiser Wilhelm, nor his later career as a much-loved Stanford professor, are of particular interest to us, but the time and the country are. Four hundred kilometres to the south, in Munich, another young scientist with similar research interests, but of a rather different stamp, had just started medical school. This was Josef Mengele.

AUSCHWITZ, 1944

The man whose name forever casts a shadow over the study of human genetics came from a well-to-do family of Bavarian industrialists. Handsome, smooth and intelligent, he refused to join the family firm and instead studied medicine and philosophy at Munich University. He was ambitious, and desired ardently to make a name for himself as a scientist, the first of his family. By the mid-1930s he had moved to Frankfurt where he became the protégé of Otamar Freiherr von Verschuer, head of another Kaiser Wilhelm Institute, but one devoted to anthropology. The dissertation that Mengele wrote there in 1935 reflects the prevailing obsession of German anthropology with racial classification and involved the measurement of hundreds of jawbones in a search for racial differences. Two later papers are about the inheritance of certain disorders such as cleft palate. All these works are dry, factual, and rather dull. They contain no hint of the young scientist's future career.

Mengele arrived at Auschwitz on 30 May 1943. He had been urged to go there by his mentor, von Verschuer, and it was von Verschuer too who had urged Mengele to take advantage of the, as it was put to him, 'extraordinary research opportunities' he would find there. By the time he arrived at the concentration camp, it contained just over a hundred thousand prisoners and the killing-machine was fully engaged.

Mengele was only one of many medical staff at Auschwitz-Birkenau, and he was not particularly senior. But after the war, it would be Mengele whom the survivors would remember. They would remember him for his physical beauty, the exquisiteness of his uniform, his charm, and his smile. They would remember him for the unfathomable quality of his personality: he was a man who could speak kindly to a child and then send it to a gas chamber. They would remember him because he was ubiquitous, and also because he was often the first German officer they saw. As the prisoners stepped from the cattle-cars onto the platform at Birkenau, they would hear him shout '*Links*' or '*Rechts*'. 'Left' and they would die immediately, 'Right' and they were spared, at least for a time.

Among those spared was a thirty-year-old Jewish woman named Elizabeth Ovitz. She and her siblings arrived at Auschwitz-Birkenau on the night of 18 May 1944. They were brought there in a cattle-car containing eighty-four other people. Weak and disoriented from the journey, the Ovitzes stood on the Birkenau railway platform under the glare of arc lights. Elizabeth asked a prisoner, a Jewish engineer from Vienna, where they were. He replied, 'This is the grave of Israel,' and pointed to the smokestacks that towered over the camp. Forty-three years later she would write: 'Now we realised everything that we knew before, and had tried to erase from our consciousness, would actually come about.' Elizabeth and her family, twelve in all, were herded to one side. It was then that they met Mengele. Surveying them with fascination he declared: 'Now I will have work for the next twenty years; now science will have an interesting subject to consider.'

The Ovitzes were Transylvanian Jews. Their father, Shimshon Isaac Ovitz, had been a scholar and Wonder-rabbi. He had a form of dwarfism called pseudoachondroplasia that leaves much of the body unaffected but causes the limbs to grow short and bowed. Rabbi Ovitz was renowned for his wisdom and compassion. Many Romanian Jews believed that, having been denied normal height by God, he was instead endowed with extraordinary and rare virtues. Amulets containing bits of parchment decorated in his finely curling Rashi script were said to have healing powers. Rabbi Ovitz had ten children of whom seven, including Elizabeth, were dwarfed. This is consistent with a diagnosis of pseudoachondroplasia, which is caused by a dominantly inherited mutation.

When Elizabeth was nine years old, her father died suddenly. His young widow, a resourceful woman, reasoned that the short stature of her children could be used to their advantage and gave them a musical education so that they could eventually form a troupe. Even as Romania and Hungary were drawn within the orbit of Nazi Germany, the Ovitz family took their 'Jazz Band of Lilliput' through the provincial towns of the fragmented and unstable states of Central Europe. In May 1942 Elizabeth Ovitz, now twenty-eight, met a young theatre manager named Yoshko Moskovitz. He was tall and handsome and besotted with her. He wrote to his sister that he had met a woman, small in size, but well endowed with talent, wisdom and industriousness. They married in November of the same year, but only ten days after the wedding Yoshko, a yellow Star of David on his coat sleeve, was drafted into a labour battalion. The couple would not see each other again until after the war. Concealing their Jewish identities, the Ovitzes continued to tour for another two years, but in March 1944 German troops occupied Hungary and, as the last and greatest of all pogroms rolled across the country, they were caught.

At Auschwitz, Elizabeth and her siblings were kept in a separate room so that they would not be crushed by the other five hundred inmates of the block; they were also allowed their own clothes and enough food to live on. For a while they were able to stay together as a family, and managed to persuade Mengele that they were related to another family from their village. They paid for survival by being given starring roles in Mengele's bizarre and frenetic programme of experimental research.

As Elizabeth Ovitz would write: 'the most frightful experiments of all [were] the gynaecological experiments. Only the married ones among us had to endure that. They tied us to the table and the systematic torture began. They injected things into our uterus, extracted blood, dug into us, pierced us and removed samples. The pain was unbearable. The doctor conducting the experiments took pity on us and asked his superiors to stop them, otherwise our lives would be in jeopardy. It is impossible to put into words the intolerable pain that we suffered, which continued for many days after the experiments had ceased.

'I don't know if our physical condition influenced Mengele or if the gynaecological experiments had simply been completed. In any event, the sadistic experiments were halted, and others begun. They extracted fluid from our spinal chord and rinsed out our ears with extremely hot or cold water which made us vomit. Subsequently the hair extraction began again and when we were ready to collapse, they began painful tests on the brain, nose, mouth and hand regions. All stages of the tests were fully documented with illustrations. It may be noted, ironically, that we were among the only ones in the world whose torture was premeditated and "scientifically" documented for the sake of future generations...'

In this, however, Elizabeth was wrong. Mengele tortured many other people as well, including a large number of twins whom he ultimately killed and dissected for the sole purpose of documenting the similarity of their internal organs. The Ovitz family walked the tightrope of Mengele's obsessions for seven months. Once, when Mengele unexpectedly entered the compound, the youngest of the family, Shimshon, who was only eighteen months old, toddled towards him. Mengele lifted the child into his arms and softly enquired why the child had approached him. 'He thinks you are his father.' 'I am not his father,' said Mengele, 'only his uncle.' Yet the child was emaciated from the poor food and the incessant blood sampling.

Mengele displayed the Ovitzes to senior Nazis. He lectured on the phenomenon of dwarfism and illustrated it with the family, who stood naked and shivering on the stage. The experiments continued until October 1944. Even as the Third Reich entered its death-throes, Mengele still brimmed with maniacal purpose, producing a collection of glass eyes from which he sought a match to Elizabeth's brown ones. As with all he did, his reason for doing so remains unfathomable.

PSEUDOACHONDROPLASIA. ELIZABETH OVITZ (1914–92), FAR LEFT, AND SIBLINGS. BAT GALIM, ISRAEL C.1949.

Auschwitz was liberated on 27 January 1945. For Elizabeth and her family the arrival of Soviet troops lifted a sentence of certain death. Nearly all of Mengele's experimental subjects were killed once he had done with them. During the following four years the family would shuttle about the wreckage of Eastern and Central Europe. Reforming their troupe, they choreographed a grim tango that they called their *Totentanz*. Each night Elizabeth, partnered by one of her brothers, would dance the part of Life to his Death.

In 1949 the family emigrated to Israel. Elizabeth Ovitz died in Haifa in 1992. Josef Mengele was never tried for his crimes, but died on a Brazilian beach in 1979.

THE BRAKE

Of the many grim ironies that the history of the Ovitz family presents us with, perhaps the greatest is that when Josef Mengele perceived that they were remarkable, he was right. People with disorders such as pseudoachondroplasia *do* tell us something important about how bones grow to the lengths that they do, and how tall we become. Mengele did not discover what this is, nor could his pointless experiments ever have told him. But half a century later it is clear that the stubby, bent and warped limbs that are the consequence of so many bone disorders speak of the phenomenon that Victor Twitty discovered: the local control of growth.

Nowhere is the dynamic nature of bone more apparent than at the ends of an infant's long bones. Each end has a region, the growth plate, from which the bone grows. Unlike the rest of the bone, which is encased in calcium phosphate, the growth plates are soft and uncalcified. On a radiogram they appear as transverse shadows that bisect the white tips of each bone. They can be seen throughout childhood and adolescence, ever decreasing in size, until by age eighteen or so they become sealed over and linear growth stops.

Each growth plate contains hundreds of columns of chondrocytes dividing and differentiating in lock step. Born at the end of the growth plate furthest away from the bone-shaft, they then swell with proteins from which they spin a cartilaginous matrix around themselves and then die. Osteoblasts march over the graves of chondrocytes, deposit calcium phosphate and yet more matrix, and at both ends the bone pushes ever further out into space.

Pseudoachondroplasia – the disorder that afflicted the Ovitzes – throws this sequence of events into disarray. The mutation occurs in a gene that encodes one of the proteins that goes into the cartilaginous matrix that chondrocytes make. Instead of being secreted, hoewever, the mutant protein accumulates in the chondrocytes, poisoning and killing them long before their time. Not all of the chondrocytes die, but the toll is enough to drastically slow growth. The result is short, bent limbs, but a torso and face that are hardly affected at all.

Pseudoachondroplasia is only one of several disorders that cause very short limbs. Another is the disorder with which it was long confused – achondroplasia itself. From Ptah-Pataikoi, dwarf deity of youth, creation and regeneration in Egypt's New Kingdom (1539–750 bc) to television advertisements for carbonated soft-drinks, there is no more common disorder in the iconography of smallness. Like its namesake, achondroplasia is caused by a shortage of chondrocytes travelling up the growth plate – but a shortage that has a very different origin.

Achondroplasia is caused by a mutation in a receptor for fibroblast growth factors. FGFs are the signalling molecules involved in the molecular clock regulating the near to the far axis of the foetal limb. After birth, however, FGFs, far from promoting the outgrowth of the limb, inhibit it.

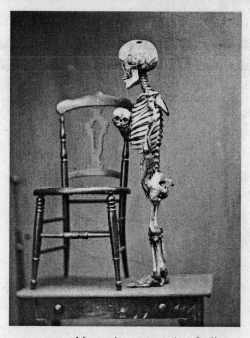

ACHONDROPLASIA. MARY ASHBERRY (D.1856) WITH THE
SKULL OF HER STILLBORN INFANT, USA.

We know this because 99 per cent of all cases of achondroplasia are caused by a mutation in which an amino acid (a
glycine) at a particular location in the FGFR3 protein sequence
(position 380) is replaced by another (an arginine). This mutation has the peculiar property of causing the FGFR3 molecule to
become hyperactive. Nearly all of the mutations discussed in this
book cause a deficiency in the quantity or efficacy of some protein, often by causing it to be completely absent. If the protein is
a signalling molecule like FGF, the disorder that we see is due to
an absence of some critical piece of information that the cells
require. The achondroplasia mutation is, however, different in

that it occasionally causes the receptor to transmit a signal into the interior of the cell even if no FGF is bound to it. The effect is like a switch that spontaneously flips on when it should be off, and that transmits a blast of unwanted information to the cells of the growing limb.

If an excess of FGF signalling causes limbs to be unusually short, then the usual role of FGFs must be to act as a brake on the growth of the infant limb. They do this by limiting the rate at which the cells of the growth plate divide. The bones of achondroplastic children have growth plates that are only a fraction of the size they should be. They contain far fewer dividing chondrocytes than those of normal children, and fewer yet that swell and form cartilage.

Achondroplasia is a relatively mild disorder. However, a surplus of FGF signalling can, in the extreme, have terrible consequences. Among the many skeletons in Amsterdam's Museum Vrolik is one that belonged to a male infant stillborn sometime in the early 1800s. When you look at the skeleton, now labelled M715, you can see quite clearly that there is something the matter with it. The child's vertebrae, ribs and pelvis are all truncated, bowed or flattened, and the skull is enormously enlarged. In his great 1849 teratological treatise, Willem Vrolik depicts the child's forehead as a large tuberose object. The stunted limbs and the large head are both characteristic of 'thanatophoric dysplasia' – death-bringing dysplasia. As the name suggests, it is fatal at birth.

Thanatophoric dysplasia is also caused by activating mutations in the FGFR3 gene, but of a far more destructive variety than those responsible for achondroplasia. The havoc they wreak shows that FGFs control the growth not only of the limbs, but of some other parts of the skeleton as well, such as the skull. The mildly domed foreheads of many achondroplastic dwarfs remind us that their disorder is a weaker version of a lethal one. Should a foetus inherit two copies of the achondroplasia mutation (by virtue of having two achondroplastic parents), it too will die shortly after birth with all the symptoms of thanatophoric dysplasia.

FGF must be only one molecule among many that limit the growth of this or that part of the body. Every organ must have devices that tell it to stop growing, and many will be unique to particular organs. There is hardly a part of the body that is not stunted or overgrown in some genetic disorder or other. Some mutations cause children to be born with tongues that are too large for their mouths; others result in intestines that do not fit inside abdominal cavities. Even muscles have their own devices for regulating growth. Belgian blues, a breed of beef cattle, are remarkable for having about a third more muscle than normal cows; their flanks resemble the thighs of Olympic weightlifters. They lack a protein called myostatin (related, as it happens, to BMPs) that instructs muscles to stop their growth. Myostatin-defective mice have about two or three times the normal muscle mass, but this gain seems to be bought at the expense of growth elsewhere, since they also have smaller than normal internal organs. Myostatin-defective people surely also exist, but there seems to be no record of them. Perhaps extra muscles are not noticed or, if noticed, are not something worth worrying about.

RENEWAL

The pseudoachondroplasia gene encodes one part of the matrix that chondrocytes spin about themselves. But it is only a minor one. Indeed, mice in which the protein has been engineered out altogether seem to suffer no ill-effects at all. One has to wonder just what it's doing there in the first place. Not so for the rest of the matrix. Most of the cartilage is made of collagen. Humans have about fifteen different types of collagens that make up about a quarter of the total protein in our bodies. Collagens are found in our connective tissue and skin. They are the stuff that holds our cells together. And they give bone much of its flexibility and strength.

Mutations that disable bone collagens cause a disorder called osteogenesis imperfecta. There are at least four forms of the disease, some of which are lethal in infancy. The most characteristic symptom of the disorder is the extreme fragility of its victim's bones. For this reason it is often known as 'glass bone disease'. The mutations have their devastating effects because of the hierarchical nature in which collagens are organised. Any given collagen protein is made up of three peptides – strings of amino acids – wrapped together in a triple helix. The triple helices are in turn grouped together in enormous fibrils that, woven together, make up the structure of connective tissue and cartilage. Each peptide is encoded by a different gene, but a single mutant gene can wreck any number of triple helices, and so any number of fibrils, and so any number of bones.

Osteogenesis imperfecta is the disorder that afflicted the

French painter Achille Empéraire (1829–98), who was himself painted by Cézanne, and the French jazz pianist Michel Petrucciani (1962–99). These artistic associations have lent the disorder, at least in France, a spurious romance (the 'glass-bone man' in Jean-Pierre Jeunet's film *Le fabuleux destin d'Amélie Poulain* springs to mind). The reality is more mundane. Children with osteogenesis imperfecta often suffer minor bone fractures of which their parents are quite unaware. When, after a more severe fracture, the children finally wind up in hospital, radiographs reveal a long history of broken and healed bones. Suspicions of child abuse often follow. In the United States, afflicted children have been taken into care by over-zealous social workers; some parents have even been jailed.

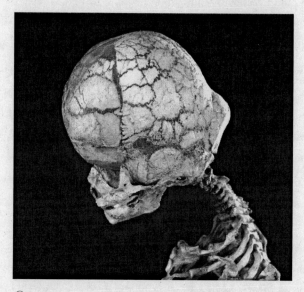

OSTEOGENESIS IMPERFECTA TYPE II. STILLBORN INFANT, AMSTERDAM.

Even once our growth plates are sealed and growth has stopped, there is no rest for the skeleton. The interiors of most adult bones are fully replaced every three or four years, while their outer peripheries, being harder, turn over about once every decade. This cycle of destruction and renewal is the product of an engagement between osteoblasts and other cells that continually wear the skeleton away, taking minute bites from its fabric and reducing it to its constituent parts, rather in the manner of so many chisels. These are the osteo*clasts*: giant cells that attach to fragments of bone and dissolve them using protein-chewing enzymes and hydrochloric acid. Bones may be built by osteoblasts, but they are carved by osteoclasts, for it is these cells that hew the ducts, channels and cavities through which nerves and blood vessels thread, and bone marrow percolates.

There are many ways to upset the balance between growth and destruction that is found in every bone. An excess of bone may be due to an excess of osteoblasts, but it can also be caused by a want of osteoclasts. Osteo*petrosis*, literally bones-like-rock, is an osteoclast disorder, the opposite of the far more familiar osteo*porosis* that is the bane of post-menopausal women. Having bones-like-rock can be lethal. There is a particularly harsh variety of the disorder that affects children and usually kills them before they turn twenty. Often they die of infections because bone accretes in the cavities where marrow is manufactured, marrow being one of the main sources of immune-system cells. Somewhat paradoxically, the bones of people with osteopetrosis also tend to fracture rather easily, the probable consequence of an architecture that has gone awry. And when fractures do occur they are not easily repaired, for among the things that osteoclasts do is to smooth away the jagged edges of our bones should we break them.

Osteopetrosis, albeit of a fairly mild variety, is thought by some to be responsible for the shortness of Henri de Toulouse-Lautrec. This is just one of several retrospective diagnoses – achondroplasia and osteogenesis imperfecta among them – that have been attempted of the French painter. None is particularly convincing, but then bone disorders are so many, and their symptoms so various and subtle, that they are easily mistaken for one another, particularly when all we know of the patient comes from biography, a handful of photographs, and a selection of self-portraits, mostly caricatures. Yet the search for 'Lautrec's disease' goes on. Part of his fascination, particularly for French physicians, comes from the fact that he was a scion of one of France's most noble houses, the Comtes de Toulouse-Lautrec, a dynasty of rambunctious southern noblemen who had, at one time or another, ruled much of Rouergue, Provence and the Languedoc, sacked Jerusalem, dabbled in heresy, been excommunicated by the Pope (on ten separate occasions) and, in the thirteenth century, felt the military wrath of the French Crown. But more than this, the impulse to diagnose Henri Toulouse-Lautrec comes from the belief that this gifted painter made his deformity part of his art.

There may be something to this. As one walks through the Musée d'Orsay in Paris or else the museum at Albi, not too far from Toulouse itself, which is dedicated to his work, what strikes you are the nostrils. In painting after painting – of the dancer La Goulue, the actress Yvette Guilbert, the socialite May

Milton, or the many other anonymous Parisian *demi-mondaines* who inhabit Lautrec's art – what we see are nostrils, gaping, dark and cavernous. It is hardly a flattering view, but perhaps it is one that would have come quite naturally to the artist, for he was rather short. By the time he was full grown, Lautrec was only 150 centimetres (four feet eleven inches) tall. Critics have also argued that Lautrec's disorder had a more subtle effect on his art: a tendency after 1893 to truncate the limbs of his models so that only the heads and torsos remain in the frame, a device for excluding that part of his own anatomy that he would much rather forget: his legs.

Lautrec's legs caused him much grief. He seems to have had a fairly healthy childhood, but by the time he was seven his mother had taken him to Lourdes, where she hoped to find a cure for some vaguely described limb problem. He was stiff and clumsy and prone to falls, and only went to school for one year, leaving when he proved too delicate for schoolyard roughhousing. By the age of ten he was complaining of constant severe pains in his legs and thighs, and at thirteen minor falls caused fractures in both femurs which, to judge from the length of time during which he supported himself with canes, took about six months to heal. He would use a cane nearly all his adult life; indeed, friends believed that he walked any distance only with reluctance and difficulty.

As he grew, Lautrec also underwent some unusual facial changes. A pretty infant, and a handsome boy, he later developed a pendulous lower lip, a tendency to drool, and a speech impediment rather like a growling lisp, and his teeth rotted

PYCNODYSOSTOSIS (PUTATIVE). HENRI TOULOUSE-LAUTREC
(1864–1901).

while he was still in his teens – traits which his parents, who
were notably good-looking, did not share. He was self-conscious
about his looks, wore a beard all his adult life, and never smiled
for a camera. Many critics have argued that it was a sort of physical self-loathing that caused him to seek and portray all that was
most vicious and harsh in his *milieu*. But then, *fin-de-siècle* Paris
could be a vicious and harsh place. One night at Maxim's, when
Lautrec had sketched some lightning caricatures of his neighbours, one of them called to him as he hobbled away. 'Monsieur,'
he said, gesturing to a pencil stub left on the table, 'you have forgotten your cane.' On another occasion, looking at one of the

many portraits he had done of her, Yvette Guilbert remarked, 'Really, Lautrec, you are a genius at deformity.' He replied, 'Why, of course I am.'

Lautrec is thought to have been afflicted by a variety of osteopetrosis called pycnodysostosis. It is caused by a deficiency in the enzyme that osteoclasts use to dissolve the protein matrix of bones. During adulthood the activity of this enzyme is partially repressed by hormones, and it is the declining levels of estrogen in post-menopausal women – and hence the unwarranted activity of the enzyme – that causes osteoporosis. Lautrec was diagnosed with pycnodysostosis in 1962 by two French physicians, Pierre Maroteaux and Maurice Lamy, but their claim has not gone unchallenged. Lautrec's most recent biographer, Julia Frey, concedes that at least some of his symptoms are consistent with the disorder, but points out that others are not. Where pycnodysostosis patients typically have soft heads – rather in the manner of the boneless Cape Malays – there does not seem to be any evidence that Lautrec's head was anything but solid.

Whatever his disorder, it seems that he shared it with several other members of his family. By the time Henri Marie Raymond, Comte de Toulouse-Lautrec-Montfa, was born in 1864 his family, though still rich, was quite inbred. The Napoleonic abolition of primogeniture had prompted an already much-reduced French nobility to keep what wealth remained in their families by the simple expedient of not marrying out of them. Henri's parents were first cousins, as were his aunt and uncle: between them they produced sixteen children, of whom four including Henri were dwarfed, the other three far more severely than he. Indeed, it is likely that at least some surviving members of that noble house still carry the mutation, though it is not likely to be expressed if they have discontinued their consanguineous habits.

Lautrec himself had no doubts about the ultimate cause of his malady. One night, in one of his favourite haunts, Montmartre's Irish and American Bar, two women were arguing about a pitiful dog whose legs shook from hip dysplasia. The dog's owner conceded that the animal wasn't handsome, but insisted nevertheless that it was pure-bred. 'Are you kidding, that dog has a pedigree? Have you taken a look at his ugly fur and his twisted feet?' laughed her friend. 'He makes you feel sorry for him.' 'You obviously don't know anything about it,' said the dog's owner, and turned to Henri who was sitting next to her. 'Tell her, Monsieur, that my dog can perfectly well be ugly and still be pedigreed.' Henri, getting down from his high barstool and standing up to his full four feet eleven inches, saluted her with a charcoal-stained hand and murmured, 'You're telling me.'

VI

THE WAR WITH THE CRANES

[ON GROWTH]

FROM THE WALLS OF THE PRADO, the Louvre and the National Gallery they stare balefully at us. As depicted by Vélazquez, Argenti, Bronzino, Carracci, Van Dyck and another dozen now forgotten painters, the court dwarfs stand clad in rich and elaborate dress, miniature daggers at their sides, surrounded by the other possessions of rich and powerful men. In one painting, a princeling stands next to a dwarf, the better to display the boy's youthful elegance. In another, a dwarf is placed next to a glossy, pedigreed hound. The man's shoulders are level with the dog's withers.

'Towards the end of the seventeenth century,' wrote Isidore Geoffroy Saint-Hilaire, 'it was necessary to dream up amusements of a special sort for the leisure of princes and it was to dwarfs that fell the sad privilege of serving as the toys of the

PYGMY DEPICTED WITH ACHONDROPLASIA. ATTIC RED-FIGURE
RHYTON C.480 BC.

169

world's grandees.' But the court dwarfs were older than that. Most of the paintings that depict them date from a century earlier. Catherine de Medici (1519–89) had set the fashion. In the hope of breeding a race of miniature humans she had arranged a marriage between a pair of dwarfs. A few years later, the Electress of Brandenburg tried the same thing, but both couples proved childless. Peter the Great took the amusement to its extreme. In 1701 he staged a wedding between two dwarfs to which he invited not only his courtiers, but also the ambassadors of all the foreign powers posted to his capital. He also ordered all dwarfs within two hundred miles to attend. A dozen small men and women rode into the capital on the back of a single horse, trailed by a jeering mob. At court some of the dwarfs, perceiving that they were there to be ridiculed, refused to take part in the fun. Peter made them serve the others.

Were all the court dwarfs unhappy, degraded creatures stripped of all human dignity? Geoffroy, writing in 1832, thought so. So had Buffon fifty years earlier. Joseph Boruwlaski, however, would not have agreed. For him, being small was a gift, an opportunity. It had lifted him out of obscurity. His *Memoirs* take up the tale:

> *I was born in the environs of Chaliez, the capital of Pokucia, in Polish Russia in November 1739. My parents were of middle size; they had six children, five sons and one daughter; and by one of those freaks of nature which it is impossible to account for, or perhaps to find another instance of in the annals of the human species, three of these children grew to above the*

middle stature, whilst the two others, like myself, reached only

that of children in general at the age of four or five years.

The Boruwlaskis were poor. Joseph was only nine years old when his father died, leaving the family destitute. Eighteenth-century rural Poland was, however, a profoundly feudal society in which patronage counted for all; Boruwlaski's mother had a patron, a young local noblewoman, the Staorina de Caorliz. Charmed by the young Joseph, she prevailed upon his mother to send the boy to live with her so that he could be educated. Boruwlaski thrived in his new home. By his early teens he was only 61 centimetres (two feet) tall, but he had acquired graces that would not have shamed the most noble of Polish youths. Things became a bit difficult for Boruwlaski when the Staorina got married and had a child, but even then he had an eye for a good thing. He became the protégé of another, even wealthier, aristocratic woman, the Comtesse de Humiecka. It was the making of him. For the Comtesse was not one to linger in the obscurity of provincial Poland; she had a yen for travel and for society. Bundling Boruwlaski into a carriage, she set out to conquer the courts of Europe.

Vienna, 1754. 'What,' asked Marie-Theresa, 'is the most remarkable thing in this room?' Boruwlaski gazed about the rococo splendours of the Schönbrunn, but knew the answer. 'The most remarkable thing in this room is the sight of a little man in the lap of a great woman.' Her Imperial Majesty, Empress of all Austria and Hungary, was delighted. In Munich, Prince Kaunitz offered Boruwlaski a pension for life. In

Lunéville the exiled Stanislaus, King of the Poles, professed himself delighted by his conversation – so much more interesting than that of his own court dwarf, an unhappy youth by the name of Bébé. The Comte de Treffan was also there, making notes for his article *Nain* in the *Encyclopédie*. In Paris Boruwlaski stayed with the Duc d'Orléans; at The Hague he had an audience with the Prince Stadholder. At Versailles the teenage Marie Antoinette gave him a diamond ring from her very own finger.

PITUITARY DWARFISM. JOSEPH BORUWLASKI (1739–1837).

Ten brilliant years passed in this manner. And then Boruwlaski fell in love. He paid his court to an actress. She rejected him with scorn. Years later he would write: 'If I can upbraid nature with having refused me a body like that of other men, she has made me ample amends, by endowing me with a sensibility which, it is true, displayed itself rather late, but, even in my constitutional warmth, spread a taint of happiness, the remembrance of which I enjoy with gratitude and a feeling heart.' But by then he could reflect on his youthful passion with calm. For he had long won the heart of another, a dark-eyed young noblewoman named Isalina Borboutin. She too had laughed at him, toyed with him, treated him like a child. But he persisted. He wrote to her, often and passionately. He petitioned the King of Poland for a pension so that he could support her. He was given one and a title as well: she relented.

Boruwlaski was a product of the French Enlightenment. In his *Memoirs* we hear the humane, rational, questing voice of the *Encyclopédistes*. 'It was easy,' he writes,

> to judge from the very instant of my birth that I should be extremely short, being at that time only eight inches in length; yet, notwithstanding this diminutive proportion, I was neither weak nor puny: on the contrary my mother who suckled me, has often declared that none of her children gave her less trouble. I could walk and was able to speak at the age common to other infants, and my growth was progressively as follows: At one year I was 11 inches high, English measure.

At three	1	*foot*	2	*inches*
At six	1	—	5	*inches*
At ten	1	—	9	*inches*
At fifteen	2	*feet*	1	*inch*
At twenty	2	—	4	*inches*

At twenty-five 2 — 11 *inches*

At thirty 3 — 3 *inches*

> *This is the size at which I remained fixed, without having afterwards increased half a quarter of an inch. My brother, as well as myself, grew till thirty years of age, and at that period ceased to grow. I cite this double proof to remove the opinion of some naturalists who have advanced that dwarfs continue to grow all their lives.*

This is fascinating and rather strange. Most people stop growing some time between the ages of seventeen and twenty. But Boruwlaski, small though he was, continued to grow throughout his twenties. It also took him a while to discover the charms of women: 'At age twenty-five I was like any lad of fifteen.' He was evidently a late bloomer.

Joseph Boruwlaski died in his sleep on 5 December 1837 in the quiet English cathedral town of Durham. He had had a happy life, a rich life. Born into obscurity, he had achieved dizzying social heights. Famed for his conversation and his skill with the violin, he had known most of the crowned heads of Europe. Ennobled by the King of the Poles, he had also won the patronage of the Prince of Wales. He could call the Duke and Duchess of Devonshire his friends. He was an ornament of Durham; its council paid him merely to live there. He had married a noble beauty, raised a family and, when he died at the distinguished age of ninety-eight, had outlived nearly all his contemporaries. It was a graceful end to a remarkable life. For Joseph, le Comte de Boruwlaski, was not merely any Continental aristocrat exiled from his homeland. He was the last of the court dwarfs.

PERFECTION IN PROPORTION

Why was Boruwlaski so small? The delay in puberty points to a possible explanation. So do several portraits in oil, half a dozen engravings, and a full-sized bronze that stands even now in the foyer of Durham City Hall. They all show that le Comte was perfectly proportioned in his smallness. True, his proportions were not quite those of a full-sized adult; they are rather closer to those of a child of the same size. But there is no sign of bone disorders such as achondroplasia or pycnodysostosis that cause limbs to grow stubby or bent. It is a kind of smallness that speaks of a failure in one of the most powerful and far-reaching molecular devices that regulate the size we are.

At the base of our brains, in a cavity of the skull, lies a gland called the pituitary. As big as a pea, it is immensely powerful. The pituitary secretes six hormones that collectively regulate the development of breasts in pubescent girls and the secretion of milk in mothers; the production of sperm in men and the maturation of ova in women; our allergic responses and the way we cope with stress.

But much of the pituitary is devoted to making growth hormone: it makes about a thousand times more of this one molecule than any of the other five. Secreted into the bloodstream, growth hormone circulates throughout the body. Its message to the body's cells is a simple one: 'grow and divide'. Growth hormone is not, of course, the only molecule that can do this. Every organ has its own molecular devices for regulating its size and shape, but the ability of growth hormone to spread throughout the body from a single source means that it simultaneously affects the growth of *all* tissues. It is the multiplier of our flesh and bones.

Joseph Boruwlaski has all the signatures of growth-hormone failure: a body the size and proportions of a four-year-old's, delayed puberty, and a briskly adult intellect. It is impossible to identify the molecular fault with any precision. A mutation in any one of half a dozen genes that control the regulation of growth hormone may have been responsible for Boruwlaski's smallness. Alternatively, he may have had lots of growth hormone, but no receptor for it to bind to. In the foothills of the Ecuadorean Andes there is an entire community of more than fifty people who have mutated receptors; when fully grown, the men are only 124 centimetres (four feet) tall. They live in just two villages and are rather inbred. Although Catholic, many of them have Jewish names; they are thought to descend from *conversos* who came to the New World in flight from the Inquisition. It is likely that they brought the dwarfism mutation with them, since exactly the same mutation has also been found in a Moroccan Jew. The Ecuador dwarfs are bright; as children they have a knack for winning prizes at school. But as they get older they tire of being teased by schoolmates and tend to drop out, and in the most recent generation not one of the adults has married.

* * *

In 1782 Joseph Boruwlaski met his physical opposite.

> *Soon after my arrival in London, there appeared a stupendous*
> *giant; he was eight feet four inches high, was well propor-*
> *tioned, had a pleasing countenance, and what is not common*
> *in men of his size, his strength was adequate to his bulk. He*
> *was then two and twenty years of age; many persons wished to*
> *see us in company, particularly the Duke and Duchess of*
> *Devonshire, my worthy protectress who, with Lady Spencer,*
> *proposed to see the giant.*
>
> *I went and I believe we were equally astonished. The giant*
> *remained sometime mute. Then stooping very low he offered*
> *me his hand, which I am sure would have enclosed a dozen like*
> *mine. He paid me a genteel compliment and drew me near to*
> *him, that the difference of our size might strike the spectators*
> *the better: the top of my head scarce reached his knee.*

Boruwlaski does not tell us the name of this man, but contemporary prints record the meeting of a dapperly dressed dwarf and a
man called O'Brien who billed himself as 'the Irish Giant'. This
hardly clarifies matters, since there were at least four 'Irish Giants'
circulating about Georgian London, two of whom called themselves O'Brien. Both O'Briens were born in Ireland around 1760
and claimed lineal descent from Brian Boru, an Irish monarch of
mythically gigantic dimensions. Both came to London in the early
1780s; one exhibited himself in Piccadilly, the other in St James's.
Both claimed they were over eight feet tall, but neither was more
than 235 centimetres (seven feet eight inches).

We know this because their skeletons have been measured. One of these men, Patrick Cotter, was buried in Bristol; his casket was found in 1906 and his skeleton examined before re-interment. The skeleton of the other, Charles Byrne, hangs in the Hunterian Museum of the Royal College of Surgeons and Physicians in Lincoln's Inn Fields, London. He is known there as Charlie, and he is an imposing sight, conveying an impression of oaken massivity. This is partly due to the brown tint of the bones, caused, it is said, by the speed and secrecy of their preparation. His jaw, chin and postorbital ridges are of a strength that must have given him a forbidding appearance in life. Towards his death,

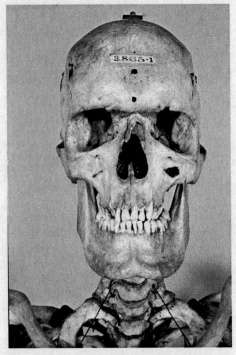

PITUITARY GIGANTISM. CHARLES BYRNE (1761–83).

which was probably due to drink, he developed the morbid fear that anatomists would seize his bones. He was right to be worried, for a contemporary newspaper describes how 'the whole tribe of surgeons put in a claim for the poor departed Irishman and surrounded his house, just as harpooners would an enormous whale'. In the event the anatomist and surgeon John Hunter got him, boiled him, and hung him where he can be seen today.

Charles Byrne had a pituitary tumor. In 1911 Sir Arthur Keith, Curator of the Hunterian, opened Charlie's skull. The indentation that had once contained the pituitary was cavernous; the gland itself must have been more the size of a small tomato than a pea. Pituitary tumors secrete vast amounts of growth hormone. They cause the cells in the growth plates of a child's limbs to divide abnormally fast, which in turn makes for super-charged growth. Childhood pituitary tumors are no less common now than when Irish giants stalked London's West End, but these days they are quickly detected and surgically removed. In May 1941, when the Hunterian suffered a direct hit from German incendiary bombs, John Hunter's giant fossil armadillo was destroyed, as were his stuffed crocodiles and many of the exquisite anatomical preparations to which he had devoted his life. Charlie, however, survived, so to speak.

PYGMIES

An old photograph shows a triptych of skeletons that used to stand in the public galleries of the Natural History Museum in London. The central skeleton once belonged to a European man.

On his left stood the hunched skeleton of a lowland gorilla; on his right, the gracefully erect one of a pygmy woman. A label, barely discernible, credits the pygmy skeleton to Emin Pasha, African explorer and Ottoman administrator. His 1883 expedition diary records that it had been unusually expensive, an outbreak of cannibalism having inflated the price of human remains in Monbuttu-land. Yet he had paid the asking price without a murmur. Pygmy skeletons were highly desirable and every museum in Europe wanted one. It had only been thirteen years since an African pygmy had first stepped out of myth and into the modern world.

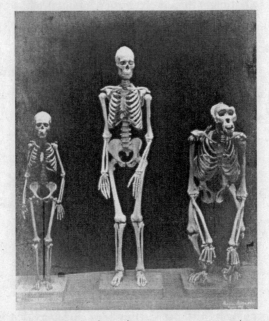

AKA PYGMY WOMAN (LEFT), CAUCASIAN MALE (CENTRE), GORILLA (RIGHT). PYGMY SKELETON COLLECTED BY EMIN PASHA, CONGO 1883.

After a few mornings my attention was arrested by a shouting in the camp, and I learned that Mohammed had surprised one of the Pygmies in attendance upon the King, and was conveying him, in spite of strenuous resistance, straight to my tent. I looked up, and there, sure enough, was the strange little creature, perched upon Mohammed's right shoulder, nervously hugging his head, and casting glances of alarm in every direction. Mohammed soon deposited him in the seat of honour. A royal interpreter was stationed at his side. Thus, at last, was I able veritably to feast my eyes upon a living embodiment of the myths of some thousand years!

The writer's name was George August Schweinfurth, a Riga-born botanist and traveller; the pygmy's name was Akadimoo. They met in 1870 on the banks of the Uele River in what is now the northernmost province of the Democratic Republic of Congo.

Akadimoo should not have existed. By the time Schweinfurth came across him the notion that there was, buried somewhere in the dark heart of Africa, a race of very small people had long been dismissed as the fancies of Greek mythographers. 'The Trojans filled the air with clamour, like the cranes that fly from the onset of winter and sudden rains and make for the Ocean Stream with raucous cries to bring sudden death to the Pigmies,' wrote Homer. Later authors wrote about a pygmy queen named Genara who had, for her beauty and her vanity, been transformed into a crane by a jealous goddess and set against her own people.

The war of the pygmies, the *Geranomachia* as the Greeks called it, is an engaging story, and one that endured for millennia. Pliny repeats and embroiders it; he places the pygmies in Thrace, Asia Minor, India, Ethiopia and at the source of the Nile, and cannot resist adding that they rode into battle on the backs of goats and were only seventy-three centimetres (two feet four inches) tall. Puzzled medieval scholastics wondered if people so small could be human, and concluded that they could not. As late as 1716 Joseph Addison wrote twenty-three Latin verses entitled *The Battle of the Pygmies and the Cranes*. Along with them he published two other Latin poems in praise of the barometer and the bowling ball. In his essay on Addison, Dr Johnson comments that some subjects are best not written about in English.

Addison's poem was the last flourish of the Homeric tradition. By the late 1600s, the hardheaded men of the Royal Society were testing legend against empirical evidence and finding it wanting. In 1699 Edward Tyson wrote a pamphlet to prove that a putative pygmy corpse he had dissected was not human. He was right, as it happens, for his pygmy was a chimpanzee. Tyson then went on to write a scathing commentary in which he pointed out that though the inhabited world was well known, no race of little men had been found; the pygmies, as well as the cynocephali (dog-headed men) and satyrs of the Greeks, were merely garbled stories about African apes.

Tyson's reasoning was clear and his intentions admirable, but he overestimated the extent of the world that was actually known. He also failed to consider that Homer's lovely simile might have been concrete knowledge transmuted. Homer certainly knew that the storks that can still be seen nesting in Greek villages in late summer, winter each year in Africa. The inference that his pygmies must live there too is plain. He was also probably remembering something distantly learned from the Egyptians. Almost a thousand years before Homer lived, Pepy II of the sixth dynasty had written to one of his generals urging him to look after a pygmy found in an expedition to the Southern Forests.

Akadimoo, the first modern pygmy, belonged to a people called the 'Aka' – a name by which they are still known. The Aka are only one of a rather heterogeneous collection of short-ish peoples who live in the African forest between the parallels $4°$ North and South. If a pygmy is defined as any member of a group with an average adult male height of less than 150 centimetres (four feet ten inches), then Africa has about a hundred thousand of them. The shortest are the Efe of the Ituri forest; their men are only 142 centimetres (four feet eight inches), their women 135 centimetres (four feet five inches). They are thought to have been there long before the invasion of the taller Bantu from the north-west about two thousand years ago.

The French anthropologist Armand de Quatrefages thought that African pygmies are the remnants of a small, dark, frizzy-haired and steatopygous people who once occupied much of the globe. This is not a ridiculous idea. In the islands of the Indian Ocean and the South China Sea there are groups of people who are almost physically indistinguishable from African pygmies. These are the 'negritos' who have been a shadowy presence in anthropology ever since the Spanish first encountered them when settling the interior of Luzon Island in the Philippines archipelago. Recent genetic studies suggest that the negritos are ancient: that they were the first Palaeolithic colonists of Asia. Like the rest of humanity, they came from Africa, but they are not especially closely related to Africans, much less African pygmies. They may have evolved smallness quite independently.

NEGRITOS. PORT BLAIR, ANDAMAN ISLANDS C. 1869–80.

Theories about the cause of pygmy shortness long antedate sure knowledge of their existence. The *Geranomachia* was a favourite theme of Attic artists, who knew only two things about pygmies: that they were short and that they did not like cranes. A red-figure rhyton from the Classical period therefore shows an achondroplastic dwarf clubbing a bird. The diagnosis of achondroplasia is unambiguous – the limbs of the bird's assailant are short and bowed, yet his torso, head and genitals are of normal size. Pompeii has yielded a fresco, now in the Naples Museum, in which bands of pygmies hunt crocodiles while others are consumed by hippos, and yet others copulate energetically on the banks of the Nile. These Roman pygmies are not deformed, but rather have the large heads and spindly limbs of emaciated three- or four-year-old children. The oddness of these images is perfectly excusable, since none of the artists had ever seen a pygmy; they were depicting the fabulous by appealing to the familiar. More surprisingly, as recently as 1960 a leading anthropologist and expert on pygmies asserted that they are small because of an achondroplastic mutation. Little is known about what makes pygmies short, but this is certainly wrong.

That pygmy proportions are not the result of any known pathology is clear from the skeleton collected by Emin Pasha. It shows that pygmies have limbs that are beautifully proportioned, but that differ from those of taller people in subtle ways. The action of natural selection over the course of tens of thousands of years has made a form more gently sculpted than the dramatic mutations familiar to the clinical geneticist. Studies of children fathered by tall African farmers on pygmy women suggest that pygmy smallness is probably not due to a single mutation, since the children have a height intermediate to that of the parents. So several genes are probably responsible for pygmy shortness. We do not know what these genes are, but we do have some idea of what they do. Careful measurements of pygmies (and thousands of them have been measured) show that compared to taller people, pygmies have relatively short legs but relatively long arms. They also have heads and teeth that are relatively large for their torsos. They have, in fact, not only the height, but also the linear proportions of an eleven-year-old British child.

By 'linear proportions' I mean the relative lengths of torso, arms and legs. Pygmy men have the broad chest and shoulders of adult men anywhere, and pygmy women have fully adult breasts and hips. But the juvenile linear proportions of pygmies immediately suggest two devices by which they should come to be so small. Perhaps they simply stop growing at age eleven. Alternatively, perhaps they grow for as long as taller people do (until age eighteen or so), but very slowly.

In principle it should be easy to distinguish between these two ways of being small – it is just a matter of measuring many pygmy children of known age to see when they stop growing. But pygmies do not know how old they are. They have no calendar and so no interest in birthdays. Occasionally, however, pygmy children have been measured. Schweinfurth traded a dog for an Aka called Nsévoué and attempted to bring him back to Europe, but they did not get far before the child succumbed to dysentery. In 1873 another attempt was made, but this time it was the explorer who died. Giovanni Mani, an Italian following Schweinfurth's trail, traded a dog and a calf for two Aka children, Thibaut and Chair-Allah, and headed north only to expire from the rigours of his journey. The children, however, went on and arrived in Rome in June 1874, where they were presented to King Victor Emmanuel II and then bequeathed, along with Mani's diaries, to the Geographical Society of Italy.

The geographers, entranced by their acquisition but puzzled what to do with it, passed the children on to Count Miniscalchi-Errizo, a Veronese nobleman. Redubbed Francesco and Luigi, they flourished under the good Count's care and were soon

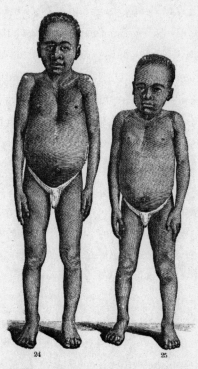

THIBAUT-FRANCESCO AND CHAIR-ALLAH-LUIGI, VERONA c.1874. FROM ARMAND DE QUATREFAGES 1895 *THE PYGMIES.*

speaking, reading and writing Italian with panache. Thibaut-Francesco taught himself piano and would pick out delicate airs though his fingers spanned less than an octave. Schweinfurth visited the boys in 1876 and recorded with delight the sight of them sauntering down the streets of ancient Verona with local friends.

The intellectual progress of the two boys was much commented on in the scholarly journals of the day, not least because it refuted the belief that pygmies might not be too bright. That this notion existed at all was partly Schweinfurth's fault. Although he had evidently been fond of Nsévoué, the published account of his travels, *The heart of Africa*, gives a rather damning estimate of his friend's ability and character. But the learned men who streamed through the Palazzo Miniscalchi to view Chair-Allah-Luigi and Thibaut-Francesco were less interested in the boys' conversation than in simply standing them against a wall and measuring them. Before they had even left Africa, Chair-Allah-Luigi and Thibaut-Francesco had been measured by at least seven scientists, and the pace picked up in Rome. The age of the boys remains in some doubt, but they were thought to be eight and twelve when they arrived in Italy, and they lived there for nearly six years. As they grew, a curious thing was noticed. They didn't have a pubertal growth spurt.

A newborn infant grows about eighteen centimetres (seven inches) in its first year. This extraordinary rate is not maintained; rather it drops smoothly, year by year, to about five centimetres (two inches) per year. At around the age of twelve for boys, ten for girls, this decline is reversed and growth rate leaps up, albeit only temporarily. Although familiar to any adolescent, the pubertal growth spurt is a rather difficult thing to measure. In 1759 the French aristocrat and friend of Buffon, Philibert Guéneau de Montbeillard (yet another count), began measuring his newly born son, and continued to do so at six-monthly intervals until the boy's eighteenth birthday. This same boy was eventually guillotined by Robespierre, but the record of his growth remains one of the most perfect of its kind. Though de Montbeillard – or rather Buffon, who wrote up the results – failed to realise it, the data show a beautiful pubertal growth spurt. At the age of thirteen, de Montbeillard's son's growth rate spiked at twelve centimetres (nearly five inches) per year. This is a very human thing. Male chimpanzees and gorillas pack on muscle at adolescence and baboons' snouts elongate, but no other primate shows this sort of skyward leap.

The pubertal spurt is driven by a burst of growth hormone. Pygmies might, then, be expected to have growth hormone levels much lower than those of taller people; but curiously, they don't. Their shortness seems to be due to a relative lack of another growth-promoting molecule called insulin-like growth factor-1, or IGF-1. As implied by its name, IGF-1 is structurally rather similar to insulin – the hormone of sugar metabolism. Growth hormone regulates the IGF gene so that levels of the two hormones in the bloodstream tend to rise and fall in synchrony. But each hormone makes a unique contribution to growth.

The proof of this is the mini-mouse. A normal laboratory mouse weighs around thirty grams when fully grown. This is rather larger than *Mus musculus* in its natural habitat (cellars, attics, barns); generations of *la dolce vita* in the world's laboratories have made the geneticist's mouse tame, slow, and slightly corpulent. Be that as it may, if a defective growth-hormone receptor gene is engineered into a laboratory mouse (rather as occurs naturally in Ecuadorean dwarfs), it grows up to be only half the size of a normal mouse. If a defective IGF gene is engineered into another mouse it grows up to about one third the normal size. If these two miniature mice are crossed, the result is the mini-mouse in which *both* genes are defective and that weighs, when fully grown, only five grams.

This, for a mammal, is minute. It is almost as small as the smallest of all mammals, the bumblebee bat of Thailand, which weighs around two grams. A British five-pence piece weighs 3.2 grams; a euro-cent 2.4 grams; a dime two grams. An adult human that was the same relative size as a mini-mouse would weigh as much as a fourteen-month-old child – a result that suggests that neither the pygmies of the Congo, nor the dwarfs of Ecuador, nor even Joseph Boruwlaski, small as they are, even begin to approach the limits of human smallness.

CRETINS

Schweinfurth's discovery set off a global hunt for other pygmies. Little people had always cropped up in explorers' logs and local myths in this or that part of the globe. Such tales had never received much credence, but in the 1890s they were assiduously collected and analysed. Suddenly there seemed to be pygmies in Guatemala, the Yucatan, the Cascade Range of British

Columbia, the Atlas Mountains of Morocco, Sicily and the Val de Ribas of Spain. An archaeologist claimed the existence of a race of Neolithic pygmies in Switzerland. Perhaps all these little people were related; perhaps they were the remnants of an earlier, shorter, version of humanity.

The fossil record shows otherwise. Our direct ancestor, *Homo erectus*, was about 160 centimetres (five feet two inches) tall; *Homo neanderthalensis* was about 170 centimetres (five feet six inches) tall; and early anatomically modern humans ('Cro-Magnon man') were only a little shorter. To be sure, there are short people in various parts of the world. Adult men of the Yanomamo tribe who live at the headwaters of the Orinoco and Amazon rivers have an average height of only 153 centimetres (five feet). The Papua New Guinean highlanders who live on Mount Goliath are also small. The enormous differences among people from around the globe show that the size we are is very malleable. We cannot be sure that smallness evolved independently in African pygmies and Asian negritos, but elsewhere in the world, smallness has evolved again and again.

I say *evolved*, but a note of caution is required. Most small people live in remote and impoverished parts of the world. It is difficult to know just how tall they would be if fed a protein- and calorie-rich supermarket diet. No one believes that African pygmies would grow much taller if transported *en masse* to California, but we would do well to remember that the children of Mayan refugees who moved to Los Angeles in the 1970s gained an additional 5.5 centimetres (about two inches) over their relations who stayed in Guatemala.

It is even possible that the most recent, and probably the last, pygmy tribe to be discovered will prove not to be pygmies at all, but rather people with a severe and rather specific nutritional deficiency. In 1954 a Burmese soldier marching through the montane forests near the joint frontiers of Burma, India, Tibet and China came across a village of small people. He was not quite the first to do so. Before Burma's independence, a series of British explorers – lean, lone Indian Army officers – had traversed back and forth across the region where the four great rivers of Asia, the Irrawaddy, Salween, Mekong and Yangtse, descend from the Tibetan Plateau. Their reports are scanty, but consistent. They record the existence of an ethnically distinct group of 'dwarfs' who seemed to have their centre in the upper reaches of the valley of the Taron, a remote tributary of the Irrawaddy. The dwarfs were variously called Darus, Nungs, Naingvaws, Hkunungs or Kiutzu. They were elusive and no one had studied them at any length, yet most accounts agreed that they were a cheerful and hardy, if notably dirty, people who tattooed their faces, lived in tree houses, and were often enslaved by the taller hill-tribes such as the Lisu. A Captain B.E.A. Pritchard measured some Nungs and found they had an average height of 158 centimetres (five feet two inches). He later drowned while trying to ford the Taron after the Nungs cut the bridge that spanned it.

In 1962 the Burmese government decided to find out more. A caravan of military men and physicians walked for two weeks across razor-backed ridges and rope-bridged ravines to the Taron Valley. Their study was published in one of the world's most obscure journals, the *Proceedings of the Burma Medical Research Society*, but it is clear and comprehensive. The Burmese found ninety-six people living in two villages. Disappointingly, there were no tree houses and no tattoos, but the men had an average height of only 144 centimetres (four feet eight inches). This was as short as the shortest African pygmies. Yet these people, who called themselves Taron after the river on which they lived, were clearly of Tibeto-Burman stock, and spoke a Tibeto-Burman language. Subsistence farmers of a meagre sort, they

DARU OR TARON. UPPER BURMA C. 1937.

lived in conditions of abject poverty and squalor. Three generations previously, the Taron said, they had crossed over from
Yunnan; a landslide had blocked the pass through which they
had come and they had been in Burma ever since.

Who were they? The Burmese weighed up the evidence and
decided that the Taron were probably identical to the Nungs of
earlier reports, and therefore a race of genetically short people.
How many more of them there were, and their precise origins,
were questions left unanswered. The hypothesis that they were
true pygmies appeared to be supported by the fact that they lived
in close proximity to taller people whose diets seemed no worse
than theirs. Yet there were disquieting aspects to the Taron. Of
the ninety-six living in the two villages, nineteen were mentally
defective. This is a high proportion, even allowing for the fact
that they were inbred (pedigrees showed many first-cousins
marriages). Several had severe motor-neuron disorders and
were unable to walk. And the Taron themselves claimed that
when they had come from China they had been of normal size;
only in Burma had they become small. That is all we know of
the Taron, and we are not likely to know more soon – foreigners have not been allowed into Upper Burma for decades. But it
is possible that the Taron are not so much pygmies, or even
dwarfs, but rather simply cretins.

It is not a pretty word, but it is the correct one. Cretins are
people who are afflicted from birth by a mix of neurological and
growth disorders. Traditionally, they have been classified into
two types: 'neurological' cretins who are mentally defective, have
severe motor-neuron problems and tend to be deaf-mute; and

'myxedematous' cretins who have severely stunted growth, dry skin, an absence of eyelashes and eyebrows and a delay in sexual maturity. A peculiarly vicious form of myxedematous cretinism, in which growth and sexual development simply stop at about age nine, is found in the Northern Congo. These Congo cretins may be in their twenties and still show no sign of breasts or pubic hair, menstruation or ejaculation, and they never grow taller than 100 centimetres (three feet three inches). This is an extreme. The Taron may have a milder form of the same disease.

Cretinism is a global scourge. In 1810 Napoleon Bonaparte ordered a survey of the inhabitants of the Swiss canton of Valais; his scientists found four thousand cretins among the canton's seventy thousand inhabitants. The location is telling. As the Taron Valley lies in the foothills of the Himalayas, so Valais lies

MYXDEMATOUS CRETINS AGED ABOUT TWENTY, WITH
NORMAL MAN. CONGO REPUBLIC 1970.

at the base of the Alps. Swiss cretins have not been spotted since the 1940s, but a belt of cretinism still tracks most of the world's other great mountain ranges: the Andes, the Atlas, the New Guinea highlands, the Himalayas. What these areas have in common is a lack of iodine in the soil. People and animals alike rely on their food for a ready supply of iodine, but in many parts of the world, especially at high altitude, glaciation and rainfall have leached most of the iodine out of the soil so that the very plants are deprived. Cretinism is caused by a diet that contains too little iodine. Globally, about one billion people are at risk of iodine deficiency; six million are cretins.

In the Gothic cathedral of Aosta, ten kilometres south-east of Mont Blanc, the choir stalls are decorated with portraits of cretins. They were carved to keep their fifteenth-century viewers mindful of the unpleasantness of Eternal Torment: a local version of the fabulous creatures and demonic creatures of misericords else-where. Many of the cretins have a curious feature: their necks are bulging and misshapen; one even has a bi-lobed sack of flesh hanging from his throat large enough to grasp with both hands. Just over a hundred years after Aosta Cathedral was built, Shakespeare would write in *The Tempest*: 'When we were boys/Who would believe that there were mountaineers/Dew-lapp'd like bulls, whose throats had hanging at them/Wallets of flesh?'

The Aosta cretins and Shakespeare's mountaineers were goitrous. Goitre is an external manifestation of an engorged thyroid, a butterfly-shaped organ located just above the clavicles. Like cretinism, it is a sure sign of iodine deficiency. When first discovered in 1611, the thyroid was thought to be a kind of support for the throat, a cosmetic device to make it more shapely. In fact it is a gland that makes and secretes a hormone called thyroxine. The thyroid needs iodine to make this hormone, and should iodine become scarce, the thyroid attempts to restore order by the rather drastic device of growing larger. The result is at first a swollen neck, then a bulging neck, and finally, in elderly people who have lacked iodine all their lives, an enormous bag of tissue that spreads from beneath the chin onto the chest, and that contains vast numbers of thyroid tissue nodules, some of which are multiplying, others of which are dying, yet others of which are altogether spent. In England this is called 'Derbyshire neck'.

A goitre is an ugly but useful thing to have, particularly for a pregnant woman. Thyroxine is yet *another* hormone, albeit not a protein, that promotes cell proliferation in the bones of foetuses and growing children. It also controls the number of cells that migrate down the growth plate to swell and die before forming bone. A foetus gets the thyroxine it needs from its mother; should it not get enough it is born cretinous. Lack of dietary iodine during childhood can also cause cretinism. And cretinism can also be, albeit rarely, a genetic disease. Many human mutations are known that disrupt the production of thyroxine, its storage, its transport around the body, or its ability to dock to its receptor.

There is also a class of mutations more vicious by far than those that simply cause thyroid malfunction. These mutations affect the pituitary. Among the hormones that the pituitary produces is one that controls the thyroid. This hormone, thyrotropin, regulates the way that the thyroid absorbs iodine, the rate at which it manufactures thyroid hormone, and the way it grows and shrinks according to need. The pituitary is the thyroid's check and its balance. Goitre is a witness to its workings. The pituitary monitors the level of thyroid hormone that circulates around the body and, should it perceive a want, begins producing thyrotropin, which then spurs the thyroid to greater efforts – in the extreme, spurs it to make a goitre. Children who have defective pituitaries are dwarfed for want of growth hormone and cretinous for want of thyroxine.

But the vast majority of the world's cases of cretinism are caused by a simple lack of dietary iodine. The tragedy of six million cretins is that the cure and the prevention of the disease is known, and costs next to nothing: it is simply iodised salt. It was the legislated spread of iodised salt in the early twentieth century that eliminated European goitre and cretinism within a generation, so that today these diseases are little more than folk-memories. Indeed, iodine deficiencies are so utterly forgotten in the developed world that outside medical and scientific circles the term 'cretin' exists only as a casual term of abuse. What is more, 'cretin' survives where comparable epithets have been justly banished from decent conversation. The word simply has no constituency, no defenders. Are the Taron of Upper Burma cretins? Is their smallness part of the vast and glorious tapestry of human genetic diversity, or are they merely victims of a peculiar form of high-altitude poverty? Were we to hear that there are no longer tribes of little people in the vertiginous gorges of the upper Irrawaddy, should we cheer or lament?

IL COLTELLO

Nearly twenty-five centuries ago, while working on a remote Aegean island, Aristotle made an observation that was at once banal, beautiful and chilling. 'All animals,' he wrote, 'if operated on when they are young, become bigger and better looking than their unmutilated fellows; if they be mutilated when full grown, they do not take on any increase of size...As a general rule, mutilated animals grow to a greater length than the unmutilated.'

By 'mutilation' Aristotle meant castration. Hence the banality of his observation that merely repeated facts as well known to any fourth-century Greek farmer as to any modern one. What makes the observation beautiful is that Aristotle thought to write it down. He has taken a barnyard commonplace, that gelded rams, stallions and cockerels are larger than intact animals, and made a scientific generalisation of it – one, moreover, that still stands. What makes these facts so chilling is that when he spoke of animals, Aristotle also meant men.

Boys who are castrated before puberty grow up to be tall, unusually so. It is a fact that is largely lost to us now, but that would have been everyday knowledge in fourth-century Athens, a city pullulating with slaves culled from all corners of the Mediterranean, among them many eunuchs. It would also have been known to any fashionable eighteenth-century Italian. The monarchs of the great opera theatres such as La Scala were not, as now, the tenors, but rather the castrati. Fêted for the range, power and unearthly quality of their voices, some castrati became rich, famous and influential. Farinelli sang for Phillip V of Spain and was given the title *Caballero*; Cafarelli became a duke and built a palazzo in Naples; Domenico Mustapha became a papal knight and Perpetual Director of the Pontifical Choir. Rossini, Monteverdi, Handel, Gluck, Mozart and Meyerbeer all wrote for them. When they sang, audiences cried '*Eviva il coltello!*' – 'Long live the knife!'– and swooned in the stalls.

CASTRATO. SENESINO SINGING HANDEL'S *FLAVIO*,
LONDON C.1723. ATTRIB. WILLIAM HOGARTH.

The Italian castrati seem never to have been measured, so we do not know exactly how tall they were. But a wealth of anecdotes and images suggests that they were taller than their contemporaries, and somewhat oddly shaped. An engraving attributed to Hogarth shows a castrato performing a piece by Handel. Mouth ajar in soaring *bel canto*, ungainly limbs akimbo, he towers above his audience. It is a caricature, and a cruel one; all the more so as the castrati suffered from much more than physical inelegance. Beyond the direct consequences of the invariably brutal surgery and the bar to marriage and fatherhood, old age frequently brought severe kyphosis, the brokenback posture that is symptomatic of osteoporosis, otherwise mostly a disease of elderly women. Many castrati also developed large and pendulous breasts. True, they never went bald, and never got prostate cancer, but these were small compensations. In eighteenth-century Italy, some four thousand boys per year lost their testicles for the sake of their golden voices. Few can ever have found the rewards that might have justified the sacrifice.

Why were the castrati so tall? Italian castrati fell from fashion and were banned by Pope Pius X in 1920; the last Vatican castrato, Alessandro Moreschi, died in 1922. But elsewhere hundreds, if not thousands, of men who had been castrated as boys survived well into the twentieth century. These were the court eunuchs, and there were many of them. At its demise, the Chinese Imperial family, last of the Qings, employed upwards of two thousand eunuchs in the Forbidden City at Peking. The last Chinese court eunuch, Sun Yaoting, was buried only in 1996 – along with his testicles, which had been carefully preserved in a jar. About two hundred eunuchs lived at the Topkapi palace in Istanbul until 1924, when the Sultan whom they had served was sent into exile, and many more must have been scattered about the vast territories once controlled by the Sublime Porte. In the 1920s some of these Istanbul eunuchs were carefully examined by a group of German physicians. What they found was distinctly odd. These elderly men, the last in a chain of eunuchs who had successively served Roman, Byzantine and Ottoman masters, had the bones of adolescents.

As children approach late adolescence, the growth plates, source of the cells that drive the growth of bones, gradually become sealed over, and it is this that finally causes growth to stop. Radiography can show how far this process has gone, and can even be used to judge the 'bone age' of a child. Where an eight-year-old has wide growth plates at each end of his long bones, those of a fourteen-year-old are narrower, while those of an eighteen-year-old are nearly, if not entirely, occluded. A handful of radiographs showed that the Istanbul eunuchs had unsealed growth plates. The inference was clear, if slightly startling: for want of their testicles, they had never stopped growing.

Testes, then, are not only the source of hormonal signals that regulate gender; they are also the source of at least one hormone that in late adolescence instructs bones to seal their growth plates and so cease growth. The nature of that hormone became apparent in 1994, when a Cincinnati clinician diagnosed a man who showed classical eunuchoid features – those disproportionately long limbs – despite having two perfectly intact and apparently healthy testes. Twenty-eight years old and 204 centimetres (six feet eight inches) in height, he was tall, but not remarkably so. What was peculiar is that, according to his driver's licence, at the age of sixteen he had been only 178 centimetres (five feet ten inches) tall. Somewhere in the intervening twelve years he had gained nearly ten inches. The molecular fault, when traced, proved to be a surprise: a mutation in the estrogen receptor.

We think of the estrogens – estradiol and estrone – as being quintessentially female hormones, and so they are. They are the hormones of breasts, periods, pregnancy and menopause. But men produce estrogens as well, and in large quantities; it is the stuff from which testosterone is made. Not all male estrogen is converted, and what remains appears to be critical in stopping the growth of bones. Two other men have been found in recent years – one in Japan, the other in New York – who cannot produce estrogens at all for want of an enzyme. They too were in their twenties and still growing fast. Conversely, children who produce an excess of estrogen tend to go through puberty very early. They grow fast but stop soon and remain short all their lives.

But there must be more to stopping growth than estrogen. Testicular estrogen may instruct adolescent bones to fuse their growth plates and so cease growing, but men without estrogen or its receptor do not grow at a rate of nearly an inch per year for their entire lives. The eunuchs who guarded Ottoman harems, regulated the affairs of the Celestial Empire and bestrode the stage of La Scala may have been imposing figures, but they were not nine feet tall. This is probably because after adolescence, bloodstream levels of IGF-1 decline, causing a general slow-down in rates of cell division throughout the body. That it does so is probably just as well.

ON BEING THE RIGHT SIZE

Growth hormone and IGF are extremely powerful growth-promoting molecules. Vital if a child is to grow to the size that it should, they must also be continually kept in check. If they are not, growth spins out of control, and the result is growth without growth's checks and balances, or, as it more commonly known, cancer.

Among the body's devices that curb IGF-1's propensity to make cells proliferate is a protein called PTEN. Infants who are born with a single defective copy of the PTEN gene show, initially, little sign that anything is wrong with them; they have, at worst, slightly larger skulls than normal. The problems come later when, as inevitably happens, the second copy of the gene mutates in a few cells of the growing child. In these cells, and their descendants, a want of PTEN protein causes cell division to spin out of control; the result is an exotic array of tumors in the mucosal lining of the mouth, lower colon, breast, ovaries and thyroid and, oddly enough, hair follicles.

These cancers are often fatal. But inheriting a defective copy of PTEN can have far more devastating consequences than this. Should the second mutation happen to occur in the first cells of the embryo (instead of in late childhood), a large fraction, perhaps even a half, of the infant's body will be completely devoid of PTEN. The afflicted fraction of the body becomes, in effect, a single, enormous and inexorably spreading tumor.

The condition is known as Proteus syndrome, named for the most versatile of Greek gods. 'Some have the gift to change and change again in many forms,/Like Proteus, creature of the encircling sea/Who sometimes seemed a lad, sometimes a lion/Sometimes a snake men feared to touch, sometimes/A charging boar, or else a sharp-horned bull,' wrote Ovid, who elsewhere calls the sea-god 'ambiguous'. The syndrome is very rare, known from no more than sixty people worldwide. Children with Proteus syndrome appear normal at birth, but their faces and limbs become increasingly distorted with age as chaotic outgrowths of bone and soft connective tissue expand over their bodies, often just on one side. They have large tracts of creased and crenulated skin, particularly on the soles of their feet, and they usually die before the age of five. In some, the cerebral hemispheres of their brains grow lopsided and they die of neural seizures; others cannot breathe because of overgrown ribs; yet others die when one of the many odd tumors to which they are prone becomes malignant. It is now believed that James

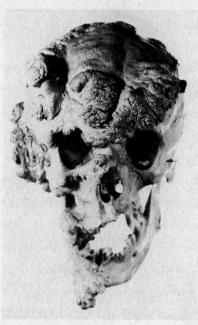

PROTEUS SYNDROME. JAMES MERRICK (1862–90).

Merrick, the so-called 'Elephant Man' who died in 1890 at the age of twenty-eight, had Proteus. If he did, then in one sense he was lucky to have lived for as long as he did.

The intimate relationship between growth and cancer is shown by dogs. A Great Dane puppy has far more IGF circulating in its bloodstream than a Chihuahua puppy does, and grows nearly eight times faster to ten times the size. It pays a cost for doing so. Great Danes, Newfoundlands, St Bernards and many other giant breeds of dogs have a risk factor of osteosarcoma or bone cancer eighty times greater than do smaller breeds. The cancer nearly always begins in one leg, and usually only amputation will prevent its spread.

Osteosarcoma is also one of the most common cancers in children. As in dogs, it usually begins in the leg bones, and then during the pubertal growth spurt when the cells of the growth plate are dividing most vigorously. In the Hunterian Museum, not too far from Charles Byrne's skeleton, is a display cabinet containing a desiccated ribcage and larynx taken from a young man who died of advanced osteosarcoma. These macabre specimens have the added horror of being covered in hard, grey nodules resembling lumps of coal. They are secondary tumors, clumps of bone-producing cells that had metastasised from the primary leg tumor – which Hunter had attempted to cure by radical amputation. In children, as in dogs, size is a risk factor for osteosarcoma. More than 50 per cent of cases are found in children who are in the seventy-fifth centile for height at any given age.

Big dogs and tall children may be more susceptible to cancers simply because they have more cells than smaller dogs and shorter children. More likely, however, it is probably not large size *per se* that is dangerous, but rather the high levels of growth hormone and IGF that big dogs and children tend to have. Pituitary tumors, of the sort that the giant Charles Byrne must have had as a child, occasionally appear in adults as well. As in Byrne's case, they produce vast amounts of growth hormone, but this doesn't cause an increase in height, since the long-bone growth plates have fused. Instead, only the bones of the jaws, hands and feet grow, a condition known as 'acromegaly'. Often the first sign that an adult has a pituitary tumor is the need for ever-larger shoes. A pituitary tumor is a moderately dangerous and unpleasant thing in itself. But it also has a nasty indirect effect, causing elevated rates of colon, breast and blood cancers (leukaemias). These cancers are not caused by metastasis of the pituitary tumor, which is benign, but rather by something it does: namely, stimulate the entire growth-hormone-IGF system.

Why this should cause high levels of cancer is not exactly clear, but one idea is that IGF stops sick cells from dying. Cells that are stricken with a potentially carcinogenic mutation often suicide. IGF overrides this altruistic impulse and so acromegalics, big dogs and tall children are relatively prone to cancers. It is as well to be clear about the magnitude of these risks. Of all the spectres that might assail a parent, childhood cancer is the least substantial. Osteosarcoma, though a pernicious and aggressive disease, is very rare: it afflicts only 1 in 300,000 children. The parents of tall teenagers should not worry; the owners of large dogs should.

The bad news for big dogs does not end there. Many people buy health insurance for their pets so that their faithful companions do not bankrupt them as, in their dotage, the dogs require heart bypass surgery. Insurance companies, seeking as ever to minimise their risk, have collected vast databases on the health and mortality rates of their clients (the dogs, not the premium-paying humans) which show quite clearly that, independent of the risk of osteosarcoma, big dogs age faster than smaller dogs. Great Danes, Newfoundlands, and St Bernards have average lifespans of four to five years; Chihuahuas and toy poodles live about ten years longer. There are about four hundred distinct breeds of dogs: for every kilogram that one of these breeds is heavier than another, it loses eighteen days of life.

These results seem to tell us that large size is generally unhealthy, but it is just a correlation. Dog breeds differ from each other in so many ways that it is difficult to attribute differences in longevity among them to differences in size alone. Ten thousand years of dog breeding is a magnificent natural experiment, but like all natural experiments, it isn't really an experiment at all – at least not in the sense of being a controlled manipulation. Fortunately, a real experiment is at hand: dwarf and giant mice. The mutant mice that are so small for want of growth hormone also live up to 40 per cent longer than their normal-sized brothers and sisters. Conversely, mice that are genetically engineered to be giant age fast and die soon. Whatever the causes of the inverse association between body size and ageing, it seems to be found in all mammals.

I am fascinated by these findings. If dogs and mice, why not people? Could it be that small people are genetically predisposed to live longer than taller people? Some scientists think so. They point to a family with defective pituitaries who live on the Adriatic island of Kruk and who seem to be (despite being both dwarfed and cretinous) rather longer-lived than the average Croatian. Or else to studies that show that the shortest American baseball players outlive taller ones by eight years. Maybe so, but the sample sizes in these studies are small, and large national surveys in Norway, Finland and Great Britain have consistently shown the opposite trend. This is hardly surprising. Socio-economic factors account for most of a population's variation in both height and health.

To be poor is to be both short and at higher risk of nearly any disease you care to name. This effect simply overwhelms any genetic tendency for the opposite trend, if such a tendency indeed exists. Will the poor always be short? Perhaps not. Young Dutchmen are, on average, 184 centimetres (six feet) tall. This makes the Dutch the tallest people in the world, taller even than Dinka goatherds or Masai *morani*. And they are getting taller: by 2012 their men will be 186 centimetres (nearly six feet one inch) tall, their women 172 centimetres (five feet seven inches). Recently, tall activists have even managed to persuade the Dutch government to raise – by twenty centimetres – the ceiling levels specified in that nation's building codes.

This supremacy in centimetres is partly because Holland is a northern country rich in cows. Being northern, it has a popula-tion that tends to be genetically rather taller than, say, their southern neighbours the Belgians, who are in turn taller than their southern neighbours, the French. Dutch children also have a high consumption of animal protein – the product of all those placid black-and-white milk cows that give the Dutch landscape its characteristic look, and the Dutch atmosphere its characteristic tang. But Holland's geographic peculiarities are probably not enough to explain the genial blond giants that can be seen in such numbers on its university campuses; for many years it has had a medical system that is excellent, efficient and egalitarian, if hard on its taxpayers' wallets, and this must surely also contribute to the general stature of its young citizens. Most remarkably, it is no longer possible to judge the socio-economic background of Dutch children from their height. Decades of social engineering have eliminated the differences that have existed there (and everywhere else) for millennia. *Égalité* has begun to reach our very bones.

But not in most countries. Elsewhere, the rich remain tall and the poor short. Young Englishmen are, on average, 176 centimetres tall, a full eight centimetres shorter than young Dutchmen. England is also an exception to the rule that northern peoples are taller than southern ones. The inhabitants of Holland may be taller than the French, and the farmers of Schezwan taller than the Cantonese, but Yorkshire man is shorter than Essex man, and the average Scot is shorter yet. Celtic *vs*. Saxon genes may make a difference, but most public health experts point to the relative poverty of northern Britain. Notoriously, the inhabitants of some especially forsaken Glasgow council estates can travel for five kilometres in any direction without finding so much as a cabbage for sale.

The poverty and short stature of the north of England's people is long-standing. More than 150 years ago, the northern cities of Leeds and Manchester became the site of the first serious investigation into the growth of British children when the social reformer Edwin Chadwick investigated the conditions of children working in the cotton mills. By modern standards, the factory children were remarkably small. Age-for-age, they were shorter than the shortest 3 per cent of modern British children, and the difference persisted, the average eighteen-year-old factory worker being only 160 centimetres (five feet three inches) tall. In 1833, the year that Chadwick published his report, the British Parliament passed a Bill against the employment of children under the age of nine.

It is precisely the antiquity of the positive association between health and height that probably accounts for the pervasive attractions of height. From George Washington to George Walker Bush there have been forty-three US presidents, and forty of them have been taller than the average American male. James Madison was famously only 164 centimetres (five feet four inches), but then he was also the architect of his nation's constitution. Presidential candidates are not taller than the people they aspire to govern simply because they are wealthier. Voters actively choose height as well: forty of the forty-three election-winners have been taller than their closest rivals. Women of all cultures seem to prefer men who are on average five centimetres (about two inches) taller than themselves. Professors, who may be expected to value the intellect above all things, behave in the same way. Full professors in American universities are on average three quarters of an inch taller than lowly assistant professors, and department chairmen are taller yet. When asked what height they should like to be,

American men of even average height invariably wish themselves taller. And who can blame them?

The pervasive attractions of height present us with a dilemma. As we learn more about the molecular mechanisms that control height, we will be able to manipulate with ever greater subtlety the size that we, or rather our children, grow to be. But what size *should* we be? The boundary between normal and pathological height is never distinct: it is a grey zone, dictated by clinical possibility, or even convenience. There are, it is true, many diseases, genetic or otherwise, of which shortness is symptomatic. But shortness, even when genetic in origin, is not always, or even most of the time, a disease. In the United States, some thirty thousand short children are currently being given recombinant growth-hormone supplements to make them grow. Most of these children are growth-hormone deficient, and for these the treatment is quite appropriate. But about a third of them have what is called 'idiopathic short stature'. That is, they are not short because they are malnourished, or because they are abused, or because they have anything identifiably, clinically, wrong with them – they are merely short. They are given growth hormone because their parents would like them to be taller.

I wonder if this is right. Giant dogs and dwarf mice suggest that growth hormone affects bodies in ways that we do not yet fully understand. Given this, it is surely neither radical nor Luddite to suggest that we should not manipulate our children's height when there is no good medical reason to do so. It is not just a matter of growth hormone either. As we learn more about the molecular devices that make us the size we are, the temptation to apply them will become ever greater. The boundary between the normal and the pathological is not only indistinct; it is mobile, ever shifting, ever driven by technology. In a way this is just as it should be. The transformation of biological happenstance into definable, curable disease is little less than the history of medicine. Should it be also this way for height? Tallness may be correlated with all sorts of desirable things, and few short men may have become President of the United States, but these are not really terribly interesting observations. Studies of short children have shown what we might have guessed: that of all the things that might affect a child's chances for happiness and success in life, height is among the least important, far less important than intelligence, health, or the quality of care the child gets from its parents. And as we mark, with pride or anxiety, the progress of our children on doorframes, it is this that we should remember.

VII

THE DESIRE AND PURSUIT
OF THE WHOLE

[ON GENDER]

IN FEBRUARY 1868, a Parisian *concierge* entered one of the rooms under his care. The room, a mere garret, dark and squalid, on the rue de l'École-de-Médecine, contained only a bed, a small table, a coal-gas stove, and a corpse. Cyan-blue skin and a dried froth of blood upon still lips showed that the stove had caused the death; the corpse itself was what remained of a twenty-nine-year-old man named Abel Barbin.

The coroner would call the death a suicide, and without a doubt, so it was. But, as the autopsy report makes clear in the matter-of-fact prose of the morgue, the death of Abel Barbin was not a tragedy compounded merely from the usual ingredients of poverty and solitude, but also from an error that had occurred thirty years before. It was the long-delayed dénouement of a

HERMAPHRODITUS ASLEEP. AFTER NICHOLAS POUSSIN 1693.

mutation that caused a single enzyme within Abel Barbin's body to fail, and fail critically, somewhere around April 1838, seven months before he was born. We know this, for we know that when the future Abel Barbin was first lifted to his mother's breast, it was not a son that she thought she held, but rather an infant girl.

This chapter is about the devices that divide the sexes and what happens when they fail; the errors that occur at that fragile moment in the life of a foetus when the events, molecule binding to molecule, take place that will decide its fate as a girl or a boy. It is about genetic mistakes that start as disorders of anatomy and end as disorders of desire. 'A man shall leave his father and his mother and cleave unto his wife and they shall be as one flesh.' Yes, but only if he is able: the biblical injunction, so blithely given, assumes so much. Not least that we know whether we should be man or wife, but that flesh will permit us to be one with another. But some of us do not know, and for some of us flesh does not permit. When it comes to sex, we are unforgiving of mistakes.

The child who would become Abel Barbin was born in Saint-Jean-d'Angély, a quiet and rather dull town on the coastal plain of the Charente, four hundred kilometres south-west of Paris. She was baptised Herculine Adélaïde Barbin, though she would call herself Alexina. She had, by her own account, a happy childhood. At least, so she would recall, years later, when writing her memoirs. She was twelve and in love:

> *I lavished upon her a devotion that was ideal and*
> *passionate at the same time.*
> *I was her slave, her faithful and grateful dog.*
> *I could have wept for joy when I saw her lower toward*
> *me those long, perfectly formed eyelashes, with an expression*
> *as soft as a caress.*

The object of her passion was an older friend, the daughter of an aristocratic family, fair and possessed of the delicate beauty and languor of the consumptive. Alexina, on the other hand, was dark, swarthy and graceless – or so she describes herself. What is more, she was from a poor family; was, indeed, a charity case in a convent school that catered to the local *haute bourgeoisie* and nobility. If the words with which she recalled her love seem to have a tinge of melodrama about them, we must remember that the language of Romanticism would have come naturally to a French schoolgirl of the 1850s; a language in which to love, to truly love, was to exalt the beloved, to abase oneself, to love without hope of return (one thinks of Stendhal's *De l'amour*); and passionate friendships were nothing strange.

Alexina was a model student, a favourite of the Ursulines who taught her. True, she would sneak to her blonde friend's cubicle at night (and, when caught, was nearly refused First Communion by her much adored and adoring Mother Superior). But such peccadilloes aside, she prospered, and at seventeen was sent to the nearby town of Le Château to train as a teacher. There, another friendship was formed, a more overtly sexual one. This was more troubling to the nuns.

From time to time my teacher would fix her look upon me at the moment when I would lean toward Thécla to kiss her, sometimes on the brow and – would you believe it *– sometimes on her lips. That was repeated twenty times in the course of an hour. I was then condemned to sit at the end of the garden; I did not always do so with good grace.*

Troubling, too, was Alexina's failure to menstruate. Her journal is allusive and shamefaced on the fact, but her meaning is clear. Cures and diets were tried to no avail. Her looks failed to cooperate as well. As classmates blossomed into rounded womanhood, she remained thin and angular. She became increasingly hairy and took to shaving her upper lip, cheeks and arms to avoid girlish taunts. Life in the convent had other torments as well. A trip to the seashore: the girls strip to their petticoats and frolic in the waves; Alexina alone watches, afraid to disrobe, rent by 'tumultuous feelings'. She has disturbing dreams.

The following year, 1857, Alexina Barbin obtained a post as assistant teacher in a girls' school and began the love affair that was to prove her undoing. The beloved was one Sara, a young schoolmistress like herself, with a cot in an adjacent dormitory. Instantly close, their friendship became one of tender attentions. Soon Sara was forbidden to dress herself; Alexina alone would lace her up – but not without planting a kiss upon a naked breast. More kisses in the oak-wood, the intensity of which Sara found puzzling but not, apparently, repugnant. Passionate outbursts followed: 'I sometimes envy the man who will be your husband!' And then, one night, Alexina won all and became her friend's lover. 'Ah well! I appeal here to the judgement of my readers in time to come. I appeal to that feeling that is lodged in the heart of every son of Adam. Was I guilty, a criminal, because a gross mistake had assigned me a place in the world that should not have been mine?'

But even as Alexina tasted the joys of requited love, rumours of the schoolmistresses' fondness for each other began to circulate. Her health was deteriorating too: 'nameless, unfathomable' pains pierced her in – we are left to infer – the groin. A doctor was summoned, and left shocked by what he found ('My God! Is it possible?'). He suggested to the school principal that Alexina be sent away, but did so in terms far too oblique for any effect. And so, happy in love, Alexina stayed. But, tormented by guilt, she confessed to Monsignor J.-F. Landriot, Bishop of La Rochelle. This elderly and worldly priest listened with compassion, and asked if he might break the seal of the confessional to consult his doctor, a 'true man of science'. And it is here that Alexina's story no longer depends entirely upon her opaquely allusive memoirs, for Dr Chesnet published:

> *Is Alexina a woman? She has a vulva, labia majora, and a feminine urethra, independent of a sort of imperforate penis, which might be a monstrously developed clitoris. She has a vagina. True, it is very short, very narrow; but after all, what is it if it is not a vagina? These are completely feminine attributes. Yes, but Alexina has never menstruated; the whole outer part of her body is that of a man, and my explorations did not*

enable me to find a womb. Her tastes, her inclinations, draw
her towards women. At night she has voluptuous sensations
that are followed by a discharge of sperm...Finally, to sum up
the matter, ovoid bodies and spermatic cords are found by
touch in a divided scrotum.

Chesnet knew well what he had uncovered: Alexina was a hermaphrodite. Medicine may have recognised hermaphrodites, but not so the law or society. A choice had to be made, and those ovoid bodies decided it. Since the seventeeth century, medical convention had held that, when gender is in doubt, gonadal sex is what matters; and Alexina had testicles. It is still so: a modern clinician would call Alexina a 'male pseudohermaphrodite', for she had only testes ('female pseudohermaphrodites' having only ovaries and 'true hermaphrodites' having both). Leaving her employment and her lover, Herculine Adélaïde Barbin shortly became, by legal statute, Abel. He appeared in public to general scandal, suffered a brief flare of notoriety in the press, and fled to the anonymity of the capital where he attempted to start life anew. And it is in Paris, just a few years later, that the memoir ends. It was found beside the bed on which he died.

GENITALS

To understand Abel Barbin and the many others whose lives have fallen, and fall, between the two sexes, would be to know all that makes us male or female. And yet his story can be simplified, reduced to its essentials. It is not merely that he fell in love with one gender rather than another, nor even that he found himself in a body whose gender was poorly suited to his desires, nor even yet that he lived in times that were unforgiving – such stories are familiar enough. No, his story is more remarkable than any of these. It is about having a body that failed to negotiate either of the two paths to gender in an altogether convincing fashion. It is, fundamentally, a story about genitals.

When we consider the male and female body we see in each, without pausing to think about it, an identity, a homology, to the other. A heart is a heart no matter which gender it sustains. Genitals are not so obvious. Their fleshy intricacies seem less versions of each other than organs of radically divergent construction which, somehow, miraculously enough, work together. Below the navel, we are mostly interested in the differences.

Anatomists, however, have other tastes. Confronted with diversity, their instinct is to simplify and unify, to search for schemes that will yoke together the most unlikely structures. This theme – the finding of homologies – runs throughout this book. But here we're concerned with something a little different: not homology among species that have long evolved apart, but rather between the two sexes. The first reasonable account of the correspondence between male and female genitals was given in 1543 by Andreas Vesalius, the founder of a great school of Renaissance anatomists at the University of Padua. Ovaries, he argued, were equivalent to testicles. And each female fallopian tube was equivalent to a male vas deferens, as was the uterus to the scrotum, the vulva to the foreskin, while the vagina, a hollow tube, was the female version of the penis itself. For Vesalius, then, female sexual organs were the same as those of males, but merely located internally. This theory seemed to explain everything. To give it maximum effect he illustrated it with a depiction of the vagina, cervix and uterus as male genitals in a state of perpetual semi-erection.

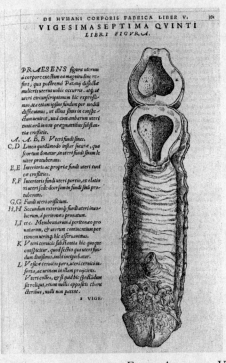

FEMALE INTERNAL GENITALIA. FROM ANDREAS VESALIUS 1543 *De humani corporis fabrica.*

Sex-education manuals invariably depict the male and female reproductive systems in some detail, often as two-toned tangles of labelled tubes. Unappetising though such diagrams may be, they are reasonably accurate. The reader who recalls one of the female reproductive tract will immediately observe that it bears little resemblance to Vesalius'. A close look at his diagram, figure twenty-seven from the fifth book of *De fabrica*, shows that it is wrong in a host of details. Vesalius showed the vagina as a long, stiff, rod-like structure, but it is not; nor does it have a swelling at its tip where the glans would be. And though the scrotum may be divided into two halves (by the raphe – the point of fusion between the two foetal genital folds), the uterus is not. Some of Vesalius' errors were surely simply made in haste. His diagram was based on the remains of a Paduan priest's mistress that had been illegally exhumed by his students, and suggests a swift and brutal dissection. Even so, the errors are puzzling; Vesalius is usually so meticulous. One cannot help but think that the vagina he drew is as much a product of what he saw on his dissecting table as it is of his theory of the unities between male and female.

That such unities – homologies – exist is certain; it's just that they're quite different from what Vesalius thought they were. But Vesalius' errors were not merely a matter of a bump more or a groove less. Allowing that it is difficult for us to see the world as a sixteenth-century anatomist did – to truly know what he did and did not know – as one looks at Vesalius' diagram, one senses that something is awry with the whole thing; that something is simply missing. Indeed, that is so. Intimate though his knowledge of the female reproductive tract was, Vesalius failed to put his finger on the most important bit of all.

It was another Paduan anatomist, Renaldus Columbus – the same Columbus who got into trouble over an extra rib – who, in 1559, discovered what Vesalius missed: the clitoris. He called it 'the Sweetness of Venus', and his description is evocative, ecstatic, and imprecise: 'Touch it even with a little finger, semen swifter than air flows this way and that on account of the pleasure even with them unwilling…When women are eager for a man', he continues, it becomes 'a little harder and oblong to such a degree that it shows itself a sort of male member.' But then he places this delightful organ in the uterus. And it is by no means clear that Columbus was really that original. Rival anatomists accused him of naming a structure that was already known to the Greeks.

Even Columbus failed to find all there was to the clitoris. In 1998, to the delight of all who have ever perceived that there is more to sex than the titillation of what is, after all, a tiny piece of flesh, the clitoris more than doubled in size. A team of Australian anatomists (headed, perhaps unsurprisingly, by a woman) working on fresh, young cadavers rather than the preserved, elderly ones that are the usual fare of medical students, revealed that the clitoris is not merely the smallish stalk of anatomy textbook and sexological dogma, but a large fork-shaped structure that surrounds the urethra and penetrates the vaginal wall.

It seemed a remarkable discovery; perhaps even the incarnation of Sigmund Freud's much derided vaginal orgasm. In Britain, newspapers hailed the discovery and wondered how so marvellous a thing could have been missed for so long. Of course, it hadn't been. The new, improved, clitoris is merely an old and well-described landmark that has been repositioned and reassessed. It is a structure long known as the vestibular bulbs, two obscure lumps of spongy tissue deeply riven with blood vessels. The seventeenth-century Dutch anatomist Jan Swammerdam thought they were part of the clitoris, but the greatest of all students of the genitals, Georg Ludwig Kobelt (author of *Die Männlichen und Weiblichen Wollusts-Organe des Menschen und Einiger Saugetiere*, 1844), considered them with care and decided, on balance, that they were not. The issue turns on whether these bulbs have the kind of rich innervation that the glans clitoris and the glans penis both have. If so, then perhaps it is reasonable to label them as part of the clitoris. Nerves there are, indeed, but it is still not clear what, if any, sort of sensations they transmit. And that is surely the critical point: an expanded clitoris that is devoid of feeling is probably unworthy of the name.

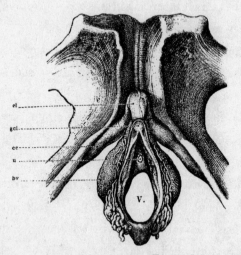

CLITORIS AND VESTIBULAR BULBS. FROM GEORG LUDWIG KOBELT 1844 *Die Männlichen und Weiblichen Wollusts-Organe des Menschen und Einiger Saugetiere.*

Large or small, Columbus's identification of the clitoris as 'a sort of male member' was accurate, if rather phallocentric. It solved part of the homology problem (clitoris = penis), but left the rest still obscure. What, then, was the female equivalent of the scrotum? And where was the male vagina? It was only in the nineteenth century that embryologists, tracing the development of the embryo's organs, truly clarified what was homologous between male and female genitals and what was unique to each.

By day 28 after conception the embryo is about half a centimetre long, has four small limb-buds, and a tail. This is when the first external signs of sex appear: nothing remarkable, just a small bump between the lower limb-buds and above the tail. The bump is the genital tubercle and it is soon surrounded by two small sets of folds, one inside the other.

The genital tubercle and its two folds together make the whole of the external genitalia. (The internal genitalia – all those tubes – have other origins.) The genital tubercle rather resembles a pale, dwarfish asparagus elongating in a flesh-coloured furrow. As it grows its fate becomes obvious: a phallus – though whether a penis or a clitoris still hangs in the balance. The innermost set of folds elongate in synchrony with the tubercle to form a kind of groove beneath it, and both – tubercle and fleshy groove – nestle within the outermost folds that are, themselves, expanding to form shallow ramparts.

This is the ground-plan of gender: the developmental events common to both males and females. The differences appear only at day 63 when the embryo is a few centimetres long and its tail is a mere docked terrier's stump. If the embryo is male, the tubercle elongates even more and the groove that runs beneath it fuses to form a tube – the 'genitourinary meatus' – the viaduct of urine and semen. Below that, the larger external folds fuse to make a sack, the scrotum – but a sack that is still empty. The testes are buried deep in the abdominal wall, and will stay there until day 160 when they migrate down to the scrotum. If the embryo is female, the genital tubercle remains much as it is, but the walls of the groove that run beneath it expand and deepen and become the labia minora. The external folds swell, but do not fuse as they do in males, and become the labia majora. The labia majora continue growing until the rather minute female genital tubercle, now more properly called the clitoris, can hardly be seen.

But what of the vagina? It has no male equivalent. Part of the vagina comes from the same folds that make the labia minora, but the innermost reaches originate from a set of ducts that also make the rest of the female sex organs, but not the male's. The homologies are now clear: penis and clitoris; scrotum and labia major; urethra and labia minora; and a vagina unique to females. Homologies that also tell us something about what happened at Saint-Jean-d'Angély around April 1838. They tell us how Abel *née* Alexina Barbin could have both a penis, albeit a small one, and a vagina, albeit a shallow one. It is as if, as a foetus, he travelled part of the route to maleness, but stopped before completing the journey. Or else that he missed the molecular signposts pointing the way.

FIFTY–FIFTY

To develop as a female is to travel a highway that is straight and wide. It is the male embryo that takes the exits; should he lose the way, he will find himself back on the route to femininity. The first signpost is the most famous of all: the Y chromosome. The Y was discovered in 1956; three years later it was identified as the master control of human gender. Within the human body, all chromosomes come in pairs matched for size and the number and kind of genes they carry. So do the sex chromosomes, but the pairing is more complex: in females, an X is paired with another X; in males an X is paired with a Y. (Normal males and females are, then, said to have XX and XY chromosomal complements respectively.) The X and the Y are physically ill-matched: the first is large, the second small. They remind one of those apparently odd couples – a large matronly woman and a small dapper man – that one sometimes finds among professionals of the Argentinean tango. The metaphor is an apt one, for no matter how implausible the pairing may seem, the two partners work smoothly together.

In sex determination, as in the tango, the Y commands and the X yields. Indeed, the proof that the Y chromosome determines sex rests upon the fact that its action is dominant over that of the X. That proof came from men who possess a Y but who also, abnormally, possess more than one X (that is, who are XXY, XXXY or even XXXXY). Such people are unambiguously male (and have perfectly normal genitalia), proving that any number of Xs are apparently powerless to curb the action of the male-making Y. True, this power does wane slightly by adolescence, when such males frequently develop breasts and an uncertain sexual orientation. But in the womb it is clear: the presence of a Y chromosome sets a foetus decisively on the route to masculinity.

The dominant behaviour of the Y is not an arrangement that should give male chauvinists any cause for delight (or ardent feminists cause for chagrin). 'Dominance' refers only to the molecular rules that are used to make sex, and not to relative male and female abilities of any sort. The dominance of the Y is also probably just an accident of history. Birds also have sex chromosomes, but they have evolved quite independently of the mammalian ones and are called W and Z rather than X and Y. These avian W and Z sex chromosomes seem to work rather like those of mammals, except that it is the female of the species that has the mismatched pair of chromosomes (WZ), while males have the matched pair (ZZ). In birds, then, the chromosome that is unique to females (W) is dominant over that also found in males (Z). It might have been so in mammals, but it seems that by chance it is not.

The search for the source of the Y's power to control gender took thirty-four years. The Y may be a dwarf among chromosomes, but it is still forty-odd million base-pairs long. Whittling this vast length of DNA down to manageable proportions took the aid of a small and unusual group of people who seemed to defy the imperious directions of the sex chromosomes. These were men who, bizarrely, seemed to lack Y chromosomes. Identified when seeking treatment for azoospermia (infertility due to immobile sperm), they were not hermaphrodites since they had testes, perfectly good male genitalia and those alone.

Geneticists call men like this 'sex-reversed females'. This sort of terminology is blatantly self-contradictory, but so is being chromosomally female yet physically male. Close inspection of the sex chromosomes of sex-reversed females showed, however, that the contradiction was more apparent than real. In each of these men, a piece of the Y had become anomalously shifted onto one of the Xs. Lacking a whole Y, they somehow had the bit that mattered. A bit that, in 1990, was found to contain a gene that encodes a transcription factor – exactly what one would expect in a gene that must directly control at least some of the many other genes that collectively make the difference between the sexes. It was named the Sex-determining Region on Y, or SRY.

The name is curious. Not sex-determining *gene*, but rather *region*. It sounds a note of hesitation, of modesty and caution. A desirable caution, since the history of the search for the Y's power is in large part a history of wasted effort. Before SRY, there was another gene, ZFY; before that, a molecule called the H-Y antigen; they were both false trails. An account of the exploration of these trails would be long and involved. Suffice it to say that the ultimate discounting of these molecules as the determinants of sex is a fine example of how science works. Hundreds of papers were published on these molecules (most especially the H-Y antigen) and many careers were built upon them, but when the evidence against them became damning, all this was set aside, and the search was started anew.

As for SRY, it was certainly a good candidate, but it was quite

possible that there was another gene, located nearby but concealed from view, that really mattered. Happily, it was not so. By
now, many lines of evidence demonstrate that SRY is the master
regulator of gender. Here is one: just as there are apparent
sex-reversed females (XX males), so too are there apparent sex-
reversed males (XY females). And, just as XX males have a
working copy of SRY where they should not, it has recently been
found that many XY females do not have a working copy of
SRY where they should. One gene, SRY, two normal states
(present on Y, absent on X), two abnormal states (absent on Y,
present on X), and a complete reversal of everything to do with
sex. It is as beautiful a demonstration of the workings of a gene
as could possibly be desired.

IF

Perhaps SRY activates a few critical genes needed for masculinity, perhaps it deactivates others needed for femininity, perhaps
both; more than ten years after its discovery, we still don't know
how it works. But we do know where it works. Dangling genitalia, hair on the chest and an adult brain excessively preoccu-
pied with sex may all be consequences of having SRY, but they
are remote ones. All that SRY does is control the fate of two tiny,
foetal organs. We know this from some experiments that were
done in Vichy France.

Foetal rabbits are all fragility; to attempt surgery upon one,
and have it survive, takes the hand of a master. Such a man was
the French biologist Alexandre Jost. In the mid-1940s, he began

a series of experiments in which he opened the womb of a pregnant dam, excised the gonads of her unborn pups, and then
sewed them all back up again. If mother and foetuses survived
surgery – his first experimental subjects often did not – he killed
them ten days later and dissected out the foetal genitalia. Jost
found that his castrati rabbits failed to develop male sexual
organs. Devoid of vas deferens, prostate, scrotum or penis, they
had instead oviducts, a uterus and a vagina. It was, perhaps, a
brutal experiment (though not much more so than the castration
of any pet), but it was also a luminous one: it showed that a male
foetus needs its gonads.

Or at least it needs its Leydig cells. These are the cells within
the gonad that make the hormones that make a mammal male.
Without SRY they and the other cells of male gonad would not
exist. One of the hormones is testosterone itself, a steroid that
Leydig cells manufacture from that much-maligned molecule,
cholesterol. The male foetus begins to make testosterone in large
quantities at around day 50, peaking at around day 150. Four
enzymes are needed to do the job; each of them represents a
signpost, which, if awry, can turn an XY foetus back to the
freeway of femininity. Mutations have been found in the genes
that encode three of them. These mutations do not cause complete sex reversal but something in between: a hermaphrodite.
More precisely, a male pseudohermaphrodite, since the testes are
there, albeit dysfunctional.

The biochemistry of masculinity is daunting. *If*, as Kipling
says, failing to keep your head while all about you are losing
theirs (etc.) will stymie male development, more surely yet will

testosterone synthesis mutations. So too will mutations that cause a failure in the differentation and growth of your Leydig cells. And so too will mutations in the testosterone receptor.

Testosterone enters the cells of the growing male foetus and binds to a protein receptor. Hormone and receptor then enter the cells' nuclei where they both bind to DNA and switch on the genes that are needed to make a male a male. Dozens of mutations in many families have been identified which cripple this receptor, usually with the effect of making the foetus completely female at birth – at least as far as external appearances go. Were the obstetrician to search carefully for a cervix and uterus, she would fail to find them. But the disorder is usually only picked up at puberty when the apparent girls fail to menstruate.

Such girls are, it is often said, exceptionally feminine – at least externally. Devoid of a testosterone receptor, they have even less exposure to masculinising hormones than do true women who invariably have some. Their one obvious masculine feature is their height: as adults they tend to be rather tall. This suggests that the difference between male and female is due to something other than testosterone (though what, exactly, is still a matter for debate). The combination of feminine looks and male height means that women without testosterone receptors are often strikingly attractive. In the 1950s at least one French woman with defective testosterone receptors made a living as a catwalk model, while a pair of receptorless identical twins from California were air stewardesses.

Many male pseudohermaphrodites are, however, not completely feminised at birth. They have what clinicians call 'ambiguous' genitalia: a phallus that is too large to be a clitoris, yet too small to be a penis. Or else a urethral opening that is located somewhere at the base of the penis rather than at the tip, yet that is clearly not a vagina. Whatever the case, nearly all would remain with the genitals they were born with were it not for surgery. But some, very few indeed, transform into their true genetic sex at adolescence.

Which brings us back to Alexina/Abel Barbin, whose growing phallus and ever increasing facial hair were the causes of such pain and confusion. There have always been, and still are, others like her. Ambroise Paré and Montaigne both met Marie Gerard, a peasant from the village of Vitry-le-François who discovered, one day while vigorously chasing swine, that she had a penis. She claimed that it merely fell out from the exertion of leaping a sty, but it is more likely that it was a gradual change, and that her – his – explanation was simply the best that suggested itself. In the event, Marie took a new name: Germain.

Alexina/Abel and Marie/Germain were both isolated cases; we do not know of any relatives or descendants of either man who showed the same symptoms. But the phenomenon of girls-who-become-men is well known in certain remote parts of the world. Until recently, the villagers of Salinas (pop. 4300), located in a remote part of the Dominican Republic, incorrectly assigned one of every ninety boys born as female, only to have them change sex at puberty. The phenomenon was so common that the villagers had a name for it: *guevedoche*, or 'penis at twelve'. On the other side of the world, in the Eastern Highlands of Papua New Guinea, something rather similar is found. There a tribe of hunters and horticulturalists called the Sambia refer to such changelings as *kwolu-aatmwol* ('changing into male thing') or, in their brutally direct Pidgin, *turnim-men*.

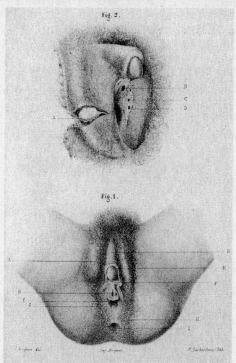

MALE PSEUDOHERMAPHRODITISM. HERCULINE BARBIN (1838–68). FROM E. GOUJON 1869 *Étude d'un cas d'hermaphrodisme bisexuel imparfait chez l'homme.*

It was the *guevedoche* who shed light on the matter; they all lack an enzyme called 5-α-reductase. In most parts of the body, testosterone is transported directly into cells to switch on genes for masculinity. But in the foetal male genitalia, much of it is first transformed into a more potent form, dihydroxytestosterone or DHT. This is what 5-α-reductase does; any mutation that cripples this enzyme will cause a lack of DHT and a failure of genital growth. DHT may be needed by the foetus, but at adolescence it is testosterone itself that does the work, and of that the *guevedoche* have plenty, so they masculinise. They can all trace their ancestry to a single woman by the name of Altagracia Carrasco; the defective enzyme is most likely her legacy. The same enzyme is, remarkably, defective in the Sambian *kwolu-aatmwol*. And although we cannot be sure, in all likelihood this was also the enzyme that Alexina/Abel lacked and which set in train the events of which I have told.

HYENAS

When I said that the route to femininity was a highway straight and wide, I meant that in humans, at least, there are few mutations that will result in a female pseudohermaphrodite, that will cause an infant with ovaries – a girl – to have masculinised genitalia. But there are some mutations that do so, and of these perhaps none has spoken more eloquently of the delicate balance of gender in the womb than a mutation that, just a few years ago, disturbed a young Japanese woman expecting her first child.

The pregnancy was without complications, or at least it was at first. But the third trimester brought the first hint that not all was well, as the young woman started growing a beard. Endocrinologists were consulted, and the cause was easily identified: instead of the estrogens common in pregnancy, her blood contained absurdly high levels of testosterone. Inevitably the child was affected as well. Though obviously female (and genetically proven to be so) the infant had, at birth, an abnormally large clitoris – about two centimetres long – and partially fused labia, sure signs of masculinisation.

It was the placenta's fault. During any pregnancy, placentas make testosterone in abundance. Normally this does not affect the foetus, for the testosterone is promptly converted into estradiol and estrone by an enzyme called aromatase in which placentas are notably rich; it was this enzyme that was defective. The mutation was a recessive one: each parent must have carried a single defective copy of the gene with no ill-effects, and each had transmitted that defective copy to their child – and to the placenta, part of which is an extension of the foetus and has its genotype. The result was a child who, in all innocence, became a kind of hormonal Trojan Horse, inflicting havoc upon her mother before she was even born.

Next to SRY itself, aromatase is arguably the single most important regulator of human gender. It sits at the crossroad of testosterone and estrogen production and directs the traffic. Girls who lack aromatase are not only born with masculinised genitalia, but as they grow up tend to be very hairy – sometimes bearded – and have enlarged ovaries. Boys who lack the enzyme, on the other hand, scarcely feel its effects – although, as discussed in the previous chapter, for want of estrogen they keep growing long after they should have stopped.

Boys, however, do not escape that easily. Not all aromatase mutations cripple the gene. There are also gain-of-function mutations that cause the enzyme to be hyperactive – and that cause an excess of estrogen and a lack of testosterone. Girls with such mutations grow up (prematurely) into short women with large breasts. Boys with such mutations are also short and, more disconcertingly, also have breasts. Aromatase mutations are not the only cause of breasts in boys and men: up to 60 per cent of adolescent boys have detectable breasts which, however, almost always disappear – at least until old age when estrogen produced by fat tissue brings them back.

People who express sex-identity mutations are often sterile. Superficial damage (ambiguous genitalia) can sometimes be repaired by surgery – though who should undergo such surgery and at what age they should do so is increasingly controversial. But all of the sex-identity mutations that I have written about have their equivalents in other mammals. Animals that find themselves between the two sexes must rarely reproduce. Such mutations, one would think, are always evolutionary dead-ends.

Always? It is difficult to generalise, for so capricious is natural selection, and so readily does it avail itself of whatever genetic variation is to hand, that the most unlikely things can happen in evolution. Spotted hyenas are unsympathetic creatures. They have ungainly bodies, cackling calls and disgusting habits. Never mind that they delight to eat carrion; they will also urinate in the water they drink and happily roll in their own vomit. More curious than this, their society is one of powerful females and milksop males. Both males and females have their own strict dominance hierarchies, but the lowliest female outranks even the most powerful male. When a clan of hyenas are having their messy way with the carcass of a wildebeest, the males have to eat quickly, for the females – which are larger – invariably drive them away. Their bulk and penchant for unprovoked aggression make female hyenas seem, one hesitates to say it, almost male. One hesitates, but then one considers their genitals.

Female spotted hyenas have genitals like no other mammal. Their most prominent feature is a clitoris as large as a male's penis, complete with a genitourinary tract, an orifice at its tip and the ability to jaunt erect during dominance displays. Beneath that, where the vagina should be, is a structure that looks remarkably like a scrotum but which contains a pad of fat rather than testicles. Lacking a vagina, the female spotted hyena copulates and gives birth through her clitoris.

And a painful business it is. The first time spotted hyenas give birth their clitoral tracts are so narrow that labour takes hours, during which time more than 60 per cent of the cubs suffocate and about 9 per cent of the mothers die. Though Aristotle denied it, the idea that spotted hyenas are true hermaphrodites persisted until the nineteenth century. They are not, for internally female and male hyenas are quite distinct – one has ovaries, the other testes and that's all. Since the females of the striped and brown hyenas (the spotted's closest living relations) have typical mammalian genitalia, spotted hyenas are, in a real sense, female pseudohermaphrodites, albeit ones in which the pathological has become normal.

The placentas of spotted hyenas have been examined; they too produce large amounts of testosterone and have – one might have guessed it – a natural deficiency of aromatase. The hyena aromatase gene has not yet been cloned, so we do not know quite how it differs from those of the brown and striped hyenas. But it is likely that the spotted hyena acquired, somewhere in its history, an aromatase gene with a mutation rather like that which our Japanese girl had. This need not be the case – it could be that the mutation occurred in another gene which regulates aromatase – but whatever the truth may be, aromatase levels almost certainly explain, at least in part, why female spotted hyena cubs are born with masculinised genitalia. What effect this testosterone has on expectant hyena mothers is difficult to say; they are hairy at the best of times.

THE OBJECT OF DESIRE

In *The symposium*, Plato gives Aristophanes a speech to account for the origin of sexual desire. There once were, Aristophanes says – he is speaking of some mythical past epoch – three human genders: male, female and hermaphrodite. These humans were not as humans are now, but rather fused together in pairs: male to male, female to female and, the hemaphrodite majority, male to female. His description suggests that these *Ür*-humans looked something like cephalothoracoileopagus conjoined twins. Able to cartwheel on their eight limbs, they were troublesome, vigorous and proud, and Zeus was puzzled what to do about them. His solution was to cut them in half. Since then, every human has sought to be reunited with his or her other half. This explains why some of us are drawn to our own sex, but most of us are drawn to the other.

The causes of our various sexual orientations are so obscure that Aristophanes' explanation is about as good as many others current today. Perhaps this is why many people resist the search for a biological explanation of sexual orientation. But unless we are Cartesian dualists (and no biologist is), the distinction between body and mind is merely a matter of the degree of our ignorance – bodies being things we understand, minds being all that we do not. And there is no doubt that the complex chain of molecular events that controls the devices of desire, our genitals, also influences its object – *whom* we choose when we give away our hearts.

What makes the case of Alexina/Abel so fascinating is that, raised as a girl, she fell in love with girls – that is, her loves were those appropriate to her sex, but her true, hidden, sex rather than her apparent one. This may seem reasonable enough, but many physicians and anthropologists think otherwise. For them, sexual orientation is made by social influences, by how a child is raised, especially early in life. In their view, Alexina/Abel having been raised as a girl among girls should have grown up to love men. Physicians take this idea from 1950s sexology; anthropologists from the writings of Franz Boas and Margaret Mead. Whatever the source, such social constructivist notions of gender are swiftly losing ground to the molecular genetic study of sexual behaviour.

Even in 1979, a study of the sexuality of the Dominican Republic *guevedoche* made a convincing case that hormones must count for something. Like Alexina/Abel, the *guevedoche* have traditionally been raised as girls and yet, almost invariably, take on a male identity as adults. They may not have had enough DHT to make good male genitals, but it seems that they received enough testosterone, either in the womb or at puberty, to make them feel like men and to make them desire and marry women. Yet they do not have it easy. As youths, they are often diffident lovers, fearing that women will ridicule them for the shape of their genitals. The label 'guevedoche' is, for the villagers of Salinas, a term used in unkindness rather than anything else.

The *kwolu-aatmwol* of Papua New Guinea have it no easier. There, only a minority of male pseudohermaphrodites are mistaken for girls; the rest are seen as boys, albeit somehow incomplete. To become men, Sambian boys must pass through an elaborate and secret set of initiation ceremonies, six in all. As in so many New Guinea cultures, these ceremonies are overtly homoerotic: young initiates must fellate older ones so that they may acquire, as the Sambia believe, a source of future semen. To the Western mind, the Sambia seem to heap cultural sexual confusion upon biological, but the boys who go through these homoerotic ceremonies mostly end up as heterosexual married men. Not the *kwolu-aatmwol*, however. They, it seems, are admitted to the lowest rungs of this ladder of initiation but not the higher ones; they may fellate, but not be fellated – indeed, their genitals would hardly permit it. And so though they may mature in body and mind, they come to be socially stranded in late adolescence, a twilight world in which they can be neither boys nor men.

VIII

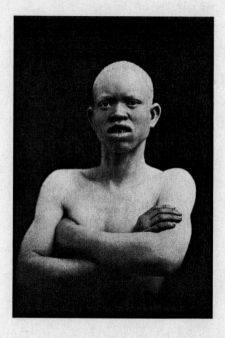

A FRAGILE BUBBLE

[On Skin]

Our species has, since 1758, borne the flattering, if not always accurate, name *Homo sapiens* – thinking Man. At least that is what the Swedish naturalist Carl Linnaeus called us in the tenth edition of his *Systema naturae*, the work which taxonomists even now accept as the first authoritative source of the names which they have given the world's creatures. It was nearly otherwise. In Linnaeus' text, directly adjacent to the word '*sapiens*' is another designation, an apparent synonym for ourselves, yet one that is somehow never explained: '*H. diurnus*' – Man of the day. It is a name that seems to have had particular resonance for Linnaeus. His notebooks show that he toyed with *sapiens* vs. *diurnus* for much of his life, and it is only in the tenth edition that the latter is firmly relegated to second place. On the

Oculocutaneous albinism type ii. Zulu man, Natal. From Karl Pearson et al. 1913 *A monograph on albinism in man.*

face of it, Linnaeus' belief that diurnality captured something special about our species is puzzling. Although we are undoubtedly daylight-loving creatures, so too are many others. It is only when one pages through the staggered typography and compressed Latin of Linnaeus' text that one finds the explanation for his diurnal dreams. Linnaeus, godfather to humanity, believed that we were not alone.

Long before palaeontologists unearthed from Serengeti dongas the bones of our extinct Hominid cousins, Linnaeus believed that the remoter parts of the world were peopled by other species of humans. He was not thinking about the humans who lived in Asia, Africa or the New World; they clearly belonged to the same species as himself. He was thinking of something altogether more exotic: a species of human that was bowed and shrunken in form, that had short curly hair rather like an African's, only fair, that had skin as white as chalk and slanted golden eyes. With eyesight as poor by day as it was acute by night, they were crepuscular, cavern-dwelling creatures who emerged at dusk to raid the farms of their more intelligent cousins. They were ancient; perhaps they had even ruled the earth before Man, but now they were on the retreat. This species, Linnaeus said, was 'a child of darkness which turns day into night and night into day and appears to be our closest relative'. True, he had never seen one, but had not Pliny and Ptolemy written of the Leucaethopes? And had they not been seen more recently, not least by his own students, in Ethiopia, Java, the Ternate Islands and Mount Ophir of Malacca? The reports seemed vivid and precise: in Ceylon they were called

Chacrelats; in Amboina, *Kakurlakos* – from the Dutch for 'cock-roach'; and everywhere they were despised. This was enough for Linnaeus, and true to his classifier's instinct, he gave them a name: *Homo troglodytes* – cave-dwelling Man. And next to that he wrote '*H. nocturnus*' – Man of the night.

What was Linnaeus thinking of? As the founder of modern biological classification, his name is second only to that of Darwin in the naturalist's pantheon. But no one reads *Systema naturae* any more, much less his many other works, and we

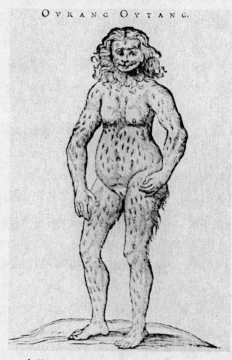

OVRANG OVTANG.

LINNAEUS' *HOMO TROGYLODYTES* OR BONTIUS'S ORANG.
FROM KARL PEARSON ET AL. 1913 *A monograph on albinism in man*.

forget that his mind was as much the mind of a medieval mystic as of an Enlightenment *savant*. Linnaeus was frankly credulous. He believed that swallows hibernate at the bottom of lakes; that if the back of a puppy were rubbed with acquavit it would grow up dwarfed; and that Lapland was the home of a creature called the *Furia infernalis*, the Fury of Hell, that flew through the air without the aid of wings and fell upon men and cattle, fatally running them through.

This last was clearly fantastic even to Linnaeus' contemporaries. Not so *Homo troglodytes*. By the 1750s it was well known that Africa at least contained creatures similar to man; Edward Tyson, after all, had dissected his 'pygmy' or chimpanzee more than fifty years previously. Another such creature, half man-half ape – the matter was all very obscure – was thought to live in the Malay Archipelago. The Dutch naturalist Jacob Bontius had illustrated just such an 'ourang-outang' in his *Historia naturalis indiae orientalis* (1658). Bontius's ourang is a fairly human, if hairy, female wearing nothing but an alluring expression; a century later Linnaeus borrowed this woodcut and relabelled it *Homo troglodytes*. Bontius himself had little to say about his ourang (though he rightly questioned the Malay belief that it was the progeny of Javanese women and the local apes), so Linnaeus grafted onto its image the ancient tradition that spoke of a remote and secretive race of unnaturally white, golden-eyed and profoundly photophobic people. It is these characteristics that yield the identity of the remainder of the mélange that is *Homo troglodytes*. Shorn of its body hair and cavernicolous habitat, it is clear that Linnaeus' Man of the night is just an ordinary human albino.

GENEVIÈVE

Linnaeus was not the only eighteenth-century naturalist with an interest in albinos. His French rival Buffon was another, but unlike Linnaeus, Buffon actually met one. In his *Histoire naturelle*, he writes of an encounter with a girl named Geneviève. She was eighteen years old, a native of Dominica, the daughter of slaves transported there from the Gold Coast, and now the servant of a wealthy Parisienne. Buffon examined her minutely. She was 151 centimetres (four feet eleven inches) tall, with slanted grey eyes slightly tinted orange towards the lens, and skin the colour of chalk. Yet her facial features, he said, were absolutely those of a *négresse noire*, a black African woman. True, her ears were stuck unusually high on her head, yet even so they were quite different from those of the *Blafards*, the albinos of the Darien Peninsula, whose ears were said to be both small and translucent. Buffon measured her limbs, her head, her feet, her hair; he devotes a paragraph to her breasts, notes that she was a virgin, and then, with interest, that she could blush.

What made Geneviève white? Buffon was certain that Linnaeus' *Homo troglodytes* was just an ape. As for the *Blafards*, *Kakurlakos* and *Chacrelats*, these were merely descriptions of anomalously depigmented people living amid an otherwise dark-skinned population. One in ten children born in the Caribbean islands, he was told, was an albino. Geneviève's parents were black, as were her siblings; whatever the cause of her whiteness it could not be contagious or even racial. Though he failed to solve the problem of albinism, when compared to the

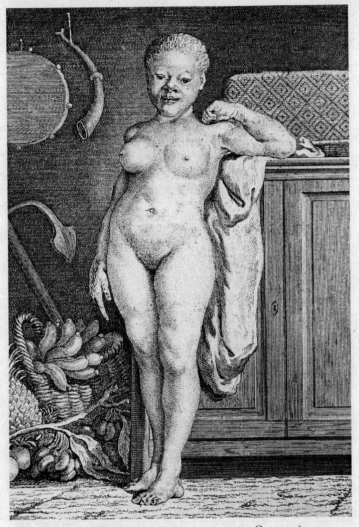

OCULOCUTANEOUS ALBINISM TYPE II. GENEVIÈVE.
FROM GEORGE LECLERC BUFFON 1777 *Histoire naturelle générale et particulière.*

fantasies of Linnaeus, Buffon left it immeasurably clearer. He also commissioned a lithograph of Geneviève, which shows her standing amid tropical fruit, quite naked and snow-white, as if in a photographic negative, smiling gently, perhaps at the absurdity of scientists.

THE PALETTE

We are a polychrome species. Yet the palette of human colour has only two pigments on it. One, eumelanin, is responsible for the darker shades in our skin, hair and eyes, the browns and the blacks; the other, phaeomelanin, for the fairer shades, the blonds and reds. As a painter mixes three primary colours to get all others, so too the various shades of our skins are given by the mix of these pigments.

Blacks have lots of eumelanin; redheads have lots of phaeo-melanin; blonds have little of either. Albinos have no skin pig-ments at all. The pigments themselves are made in cells called melanocytes that are found within the top layers of the skin, the epidermis. These melanocytes package pigments into sub-cellular structures called melanosomes which they then transfer to the skin cells immediately above them, giving them colour. Mutations in several genes cause albinism. The most common disables one of the enzymes that melanocytes use to make pigment. In such cases even the eyes are devoid of pigment, and their redness comes from the retina's blood vessels. The absence of pigment makes albinos sensitive to light and they often squint – hence the photo-phobia and slanted eyes of the *Kakurlakos* and *Blafards*. But some albinos have at least some pigment in their eyes, and in these cases the defect lies in a protein that is called, somewhat enigmatically, 'P', used in the packaging and transport of melanosomes. Geneviève's eyes were grey, not red, and it is almost certain that both copies of her P gene were defective. We can even guess what the mutation was. The most common cause of albinism in Africa is homozygosity for a 2.7 kilobase-pair deletion in the P gene. The same mutation is found in the Caribbean and among blacks in the United States as well, carried there by the slave trade.

There are no tribes, races or nations of albinos anywhere in the world; however, Pliny's Leucaethopes are not entirely without foundation. About 1 in 36,000 Europeans is born albino, and 1 in 10,000 Africans. But the number jumps to 1 in 4500 among the Zulu and 1 in 1100 among the Ibo of Nigeria, and in very local populations the frequency can become even higher. In 1871, *en route* to his encounter with the Aka pygmies, George Schweinfurth came across some.

> There is one special characteristic that is quite peculiar to the Monbuttos. To judge from the hundreds who paid visits of curiosity to my tent, and from the thousands whom I saw during my three weeks sojourn with Munza, I should say that at least 5 per cent of the population have light hair. This was always of the closely-frizzled negro type, and was always associated with the lightest skin that I had seen since leaving lower Egypt...All the individuals who had this light hair and complexion had a sickly expression about the eyes and presented many signs of pronounced albinism.

That albinism can be so common is a bit surprising. African albinos have, by any account, a hard time of it. Not only do they often suffer social discrimination and have difficulty finding marriage partners, but for want of pigment they cannot work for any length of time outdoors, and they are also prone to melanomas, a particularly destructive variety of skin cancer. These selective disadvantages should act to keep albino genes, and hence albinos, rare. Some geneticists have suggested that one reason for the high frequency (1 in 200) of albinism among the Hopi Indians of Arizona is that albino men, excused from working in the fields, stay at home and therefore dally among the women. But the evidence for this seems to rest on the charms of one old Hopi gentleman who was reputed to have fathered more than a dozen illicit children.

PIEBALDS

Those children would have fascinated Buffon. In his search for an explanation for albinism, grasping at a theory of inheritance that did not yet exist, he was keen to know what the offspring of a union between an albino and someone with normal pigmentation would be. He thought they might be piebald. In the *Histoire naturelle* he gives another lithograph. This one is of a girl, perhaps four years of age, standing amid a clutter of exotic artefacts: a parasol, axes, a blanket and a feathered headdress. A small parrot, a household pet, perches upon her hand suspended in mid-air. The girl has a two-tone body: a mosaic of black and white.

PIEBALDING. MARIE SABINA, COLUMBIA 1749. FROM GEORGE
LECLERC BUFFON 1777 *Histoire naturelle générale et particulière.*

Buffon never met the child, knew little about her origins, and described her entirely from a picture. Painted in Columbia by an unknown artist around 1740, the portrait was dispatched to Europe on a Spanish vessel which was promptly seized by the West Indies squadron of the Royal Navy. Now a trophy of war, the picture was taken to Carolina where it was copied at least twice. One of these copies, or perhaps the original, was sent to London, but this ship was plundered as well – it was the French navy's turn – and the painting was placed in the hands of the Burgomaster of Dunkirk, a M. Taverne, who sent it to Buffon. And so the War of the Spanish Succession brought Marie Sabina, the piebald child, to the eyes of Europe's greatest naturalist.

Buffon was enchanted. His copy of the portrait, which is now lost, bore the following inscription:

The True
Picture of Marie-
Sabina who was born
Oct 12 1736 at Matuna a
Plantation belonging to
the Jesuits in the City of
Cartegena in America of
Two Negro Slaves named
Martianiano and Patrona.

In a letter that he sent with the portrait Taverne wrote: 'In spite of the legend, I think that the child is the issue of a union between a white and a *négresse*, and that it was to preserve the honour of both the mother and the Society [of Jesus] whose slave she was, that it states that both the parents were black.' Buffon replied that although he initially thought that Taverne's explanation might be true, upon reflection he doubted that it could be. There were thousands, millions, of people of mixed black and white blood, and they all appeared to be uniform brown in colour. Perhaps, he continued, the child was the progeny of a black and an albino – one of those anomalous *Blafards*. And that is all we know about Marie Sabina, bar a brief mention by the Jesuit geographer and ethnographer José Gumilla, who in his *Orinoco illustrado, y defendido, historia natural y geographica de este gran rio* (Madrid, 1745) records that he encountered her as an infant in a plantation hospital, told her mother (who was recuperating) to beware that others did not cast an evil eye upon her daughter, and concluded that the child's peculiar appearance could almost certainly be blamed on the dog, a household pet, which had the misfortune to be spotted as well.

Buffon's hypothesis – that piebald children were the progeny of albinos and blacks – ran for nearly two hundred years. It was certainly a more reasonable theory than Gumilla's spotted dog, yet its longevity remains surprising, since in that time at least four other piebald children emerged from the Caribbean onto the pages of learned journals, and not one had an albino parent. Besides Marie Sabina, there were John Richardson Primrose Bobey (b.1774, Jamaica), Magdeleine (b.1783, St Lucia), George Alexander Gratton (b.1808, St Vincent), and Lisbey (b.1905, Honduras). Each child was celebrated in its day. Portraits of

Marie Sabina now hang in Williamsburg, Virginia, and at the Hunterian Museum in London; Magdeleine has a statue at Harvard University; and in Marlow, Surrey, George Gratton has a grave with the epitaph 'Know that there lies beneath this humble stone/a child of colour haply not thine own.'

The most recent of these Caribbean piebalds, Lisbey, featured in an article written by the British geneticist Karl Pearson in 1913. Like Buffon, Pearson thought that piebalding had something to do with albinism. He does not suggest that the child's mother had an affair with an albino – a photograph of the family shows a lace-clad matriarch of seemingly imperturbable

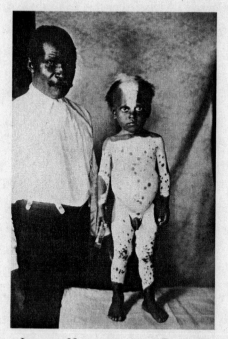

PIEBALDING. LISBEY, HONDURAS 1912. FROM KARL PEARSON ET AL. 1913 *A monograph on albinism in man.*

moral rectitude. Instead he questions Lisbey's ancestry, postulating the existence of an albino forebear. Pearson's hypothesis was a bit more complex than this, for he also proposed that an albino ancestor will only cause piebalding when one of the parents is particularly dark – and Lisbey's father was, in Pearson's words, 'a coal black negro'. It is a convoluted explanation and one that is difficult to understand from a modern point of view. We now know that piebalding has nothing to do with albinism but is instead caused by dominant mutations in an altogether different set of genes, and that these mutations can occur in people of any colour – not to mention horses, cats, and a strain of mouse called 'splotch'. They are no less fascinating to us than Marie Sabina was to Buffon. Among other things, they tell us about the strange origin of the cells that colour our skins.

Melanocytes spend their lives in the skin, but they are immigrants there. Where most of the skin is ectoderm, melanocytes are the products of a tissue called the neural crest. At about day 28 after conception, neural crest cells flow out of the newly formed dorsal nerve cord and pour themselves around the foetal head to make the face. But some neural crest cells travel much further than this. As a river fans out over its delta, streams of neural crest cells course down from the escarpment of the dorsal nerve cord and penetrate to the embryo's farthest reaches. In one part of the body they form nerves, in another muscles, yet elsewhere they invade developing glands. And some become melanocytes which, early in foetal life, invade the lower layers of the skin where they settle down to produce pigments. Neural crest cells make our faces, and they also lend them colour.

Molecular devices are required to make a naive neural crest cell form a melanocyte rather than some other kind of cell, and also to guide the melanocytes to their final destinations. Mutations in at least five distinct genes cause piebaldism, and each of them disables one or more of these devices, so causing patches of skin that are devoid of melanocytes and therefore perfectly white. Some piebalds have only a white forelock, some have bodies covered in patches, and some have eyes of different colour. Yet others have more serious disorders. A few piebald infants have a debilitating intestinal condition known as 'mega-colon' – a lower intestine that is swollen with massive constipation caused by the absence of gut nerves that drive defecation. These nerves too have their origin in the neural crest. Piebald children are also prone to deafness, for it seems that melanocytes serve some critical function in the inner ear.

DINKA VS. DUTCHMAN

Guinea pigs, dogs, cats and cattle may have been bred for variety of colour, but only humans come naturally in so many different shades. What gives us our skin colours? It is a curious thing, but for all that geneticists have learned about the causes of abnormal pigmentation, they have yet to give an account of the genes responsible for the difference in skin colour between, say, a Dinka and a Dutchman.

Why is this? In part it is due to the sheer difficulty of the problem. Geneticists agree that more than one gene makes the difference between naturally dark and fair skin (were it only one gene, we would know it by now), but beyond that the guesses range between two and six, interacting in complex combinations to give any particular shade of pink, tan, brown or black. This makes things difficult. When many genes, each of which has many variants, combine to affect some property of the human body, the molecular identification of those genes becomes a challenging exercise in applied statistics. When the property in question is a disease – heart disease or non-insulin-dependent diabetes – geneticists have embraced the challenge with a will. They have been more cautious about studying skin colour.

This is understandable. Ever since Linnaeus divided the world's people into four races – *Asiaticus*, *Americanus*, *Europaeus*, *Afer* – skin colour has been misused as a convenient mark of other human attributes. Linnaeus distinguished his four races not only by the colour of their skins but also their temperaments: *Asiaticus* was 'stern, haughty, avaricious and ruled by opinions'; *Americanus* 'tenacious, contented, choleric and ruled by habit'; *Afer*, seemingly devoid of any redeeming virtue, was 'cunning, slow, phlegmatic, careless and ruled by caprice'. What of his own race? *Europaeus*, Linnaeus thought, was 'lively, light, inventive, and ruled by custom'. This was the beginning of an intellectual tradition that, via the writings of Arthur, Comte de Gobineau, the nineteenth-century theorist of Aryan supremacy, culminated in the most systematic chromatocracy that the world has ever known: apartheid South Africa.

For nearly half a century the architects of the South African *laager* held the world at bay and devoted much of that nation's abundant resources towards the hopeless task of dividing the racial seas. In the endless negotiations as to who could or could not sit on park benches marked *net blankes* (whites only) every policeman, magistrate, employer, practically every citizen, became an expert on racial identity. South African law was always deliberately vague as to what made someone *blank*, *swart* or a *kleurling* ('coloured' – in apartheid's parlance, someone of mixed African and European descent). In part it was just who you knew, where you came from, what people thought you were. But mixed in with these social criteria was an elaborate array of pseudo-scientific tests that would, their proponents declared, infallibly betray African ancestry in someone trying to 'pass for white'. Some placed their faith in the 'pencil test' – predicated on the notion that a pencil stuck in someone's hair would only remain there if the subject was at least part black. Others held that the colour of the skin beneath the fingernails was critical, or else spoke of knowledgeably of eyelids and the Mongolian Spot. Yet others appealed to the colour of the genitals ('the scrotum test'). In the segregation of schools, hospitals, jobs, indeed every aspect of public life in South Africa between 1948 and 1990, the destiny of a child could turn on the precise shade of almost any of his or her body parts.

In 1973 a forty-year-old Cape Town housewife named Rita Hoefling, who had until then enjoyed the privilege and security that came with being a white South African, began to turn black. She had been diagnosed with Cushing's disease, a disorder caused by hyperactive adrenal glands. The glands were removed and for a while all seemed well, until she noticed that her skin was becoming rather dark. It wasn't just a matter of a tan, but rather a deep bronze, that altered her whole appearance – indeed, made her look like a *kleurling*.

The first humiliations were small ones, the stuff of 'petty' apartheid. Thrown off a 'whites only' bus by a zealous conductor, she was forced to carry a card that explained and excused her dark skin. But in apartheid South Africa any citizen could be a self-appointed Race Commissioner, and it was not long before Rita felt compelled to move to another area – only to have her new neighbours issue a petition of protest as well. All this in Cape Town, even then South Africa's most cosmopolitan and racially tolerant city. The strain eventually told on her family. When her father died, Rita was not allowed to attend his funeral: 'I do not want,' said her mother, 'to be embarrassed by your black body at Daddy's grave.'

Driven from the white community, Rita Hoefling was befriended and sustained by blacks. They welcomed her into their homes in the segregated townships, and kept her sane. She became fluent in Xhosa. And then, one day in 1978, Rita spontaneously turned white again. She attempted to return to her old life, but by then her husband – a former Royal Navy officer – and her children had left her. For the last ten years of her life she lived on charity and a small pension and moved between grimy bedsits in Cape Town's slums. It was in such a bedsit that in 1988, aged fifty-five, she died of bronchial pneumonia.

Rita Hoefling had a disorder called 'Nelson's syndrome' which occurs in about a third of patients who have adrenodectomies. One of the critical tasks that the adrenal gland does (rather like the thyroid) is keep the pituitary gland in check. In the absence of her adrenals, Rita's pituitary began to grow, became tumorous, and produced a surplus of pituitary hormone that caused her skin to darken.

Charlie Byrne, the Irish Giant, also had a pituitary tumor. It may seem surprising that a tumor in a single organ can manifest itself in such different ways, but the pituitary is a remarkably versatile organ. Not so much a hormonal factory as an industrial park, each of the half-dozen-odd hormones that it secretes is the product of a group of specialist cells. This means that tumors that start in different pituitary cell types can have very different consequences. Most pituitary tumors start in cells devoted to producing growth hormone and so cause either gigantism or acromegaly. More rarely the tumor starts in cells devoted to the production of a group of hormones called melanotropins.

Like growth hormone, melanotropins circulate throughout the body; however, where growth hormone affects nearly all of the body's cells, melanotropins tend to be more discriminating. Among the cells they affect most spectacularly are the melanocytes. When the hormone binds to its receptor on the melanocyte, the cell begins to produce eumelanin, the pigment that gives us the dark shades in our skin, hair and eyes. Just as too much pituitary growth hormone causes the over-multiplication of flesh and bone, an excess of melanotropin causes our skin to bronze – at least it does in fair-skinned people. But melanotropins do not simply turn eumelanin production on. Children who have no melanotropins are not blonds, but redheads. And they are fat.

They are fat because one of the melanotropins, a molecule called α-melanocyte-stimulating hormone (α-MSH), does more than its name suggests. On melanocytes it binds to, and activates, a molecule called melanocyte-stimulating hormone receptor-1 or MC1R. In the brain, however, it binds to another receptor called MC4R that is encoded by a different gene. The brain receptor controls appetite. When α-MSH activates MC4R a neuronal signal tells us to stop eating. Children who lack α-MSH are obese because they simply do not know when to say to say 'when'.

Yet not all redheads are fat. Indeed, casual recollection suggests that rather few are. Why is this? The answer appears to be that most redheads do not owe their fiery locks and translucent skins to a lack of any hormone but to unusual receptors. When MC1R is active, melanocytes make eumelanin – brown and black pigments; inactive, they make phaeomelanin – red pigments. Red-haired Celts have receptors that are more or less permanently inactive – something they share with red setters, red foxes, and red-haired Highland cattle.

Note the weasel-word – 'unusual'. Throughout this book, I have used the language of clinical genetics. I have spoken of 'mutations' that 'disable' proteins or else render them 'defective'. But there are so many redheads around that it seems a bit harsh to speak of their genes in this invidious fashion. And yet the question niggles: are redheads mutants?

Whether a given genetic sequence is a mutation rather than a polymorphism hinges on two issues: its global frequency and its usefulness – mutations being rare and harmful, polymorphisms being generally neither. As far as frequency goes, redheads may be common in northern Europe (6 per cent in Aberdeen), but globally they are rare. Worse, a count of heads overestimates the frequency of the 'redhead gene'. This is because each redhead is unusual in his or her own way. MC1R comes in at least thirty different versions, and many of them are found in Ireland. Six, but perhaps as many as ten, of these human MC1R versions, in a multiplicity of combinations, cause red hair – be it auburn, deep red, orange or strawberry blond. Africans, by contrast, all have just one kind of MC1R.

Globally, any single red hair version of the MC1R gene is so vanishingly rare that we must, it seems, call it a mutation rather than a polymorphism. But perhaps an argument can be made for utility? Some have speculated that northeners need lighter skins in order to garner sunlight for the manufacture of vitamin D, without which they would suffer rickets, a bone deformity. Darwin thought that the variety of human colour was due to sexual selection – generations upon generations of perfectly arbitrary choice for beautiful mates. This is a pleasing but difficult-to-prove hypothesis – at least if we discount Henri Toulouse-Lautrec's belief that redheads give off an especially erotic odour.

On the other hand, it is easy to make a case against the usefulness of red hair. The uniformity of MC1R in Africa tells us that dark skin is needed in the tropics – there is no doubt that it protects against skin cancer. Removed from soft northern light, redheads are easily ravaged by the sun. Their MC1R genes give them delicate complexions that refuse to tan but only burn. Many Australian children are descended from Scottish and Irish immigrants, and Australian law ensures that they all wear hats and long sleeves in their schoolyards. None of these arguments is conclusive. But the evidence tends to suggest that, delightful though it may be to look at, red hair is not good for anything at all. MC1R in northern Europeans may simply be a gene that is decaying because it is no longer needed, rather as eyes decay in blind cave-fish.

THE STORY OF PETRUS GONSALVUS
AND SHWE-MAONG

Pale, and proud of it, nineteenth-century European anthropologists typically ordered humanity by skin colour. Perhaps unsurprisingly, scholars from elsewhere have often seen matters differently. Upon returning from a European tour, the Chinese *savant* Zhang Deyi (1847–1919) informed his compatriots that many Frenchwomen had long beards and moustaches. Eschewing the skin-colour geographies of their European counterparts, Chinese anthropologists made maps showing which of the world's people were or were not hairy. They were fascinated by the Ainu, a relatively hirsute northern Japanese people whom they depicted as a race of dwarfish ape-men. The Ainu are, of course, nothing of the sort. True, Ainu men take a traditional pride in the length of their locks and beards (neither of which they trim), yet they have no more body hair than most Europeans. But then, learned Qing commentators also compared European visitors to macaques, a pleasant tradition that persists in Singapore, where foreigners are still called *angmo* or *angmogao*, Hokkein for 'red-haired ape'.

It is perhaps not quite fair to single out the Chinese for their preoccupation with hair (it was, after all, almost certainly a white South African who invented the 'pencil test'). And Europeans may be hairy, but this has never made them especially sympathetic to people who are hairier yet. Several genetic disorders called hypertrichoses cause infants to develop lush growths of hair on their noses, foreheads, cheeks and ears, limbs and torsos – parts that are, in most babies, only modestly clad. Grown up they have been the wild men, *Waldmenschen* and *femmes sauvages* of early travellers; the *hommes primitifs* and *Homo hirsutus* of taxonomists, and the dog-bear-lion-ape-people of fairground hucksters.

In the collection of the Capodimonte Museum in Naples there is a painting by Agostino Carracci, elder of the Bolognese artist brothers. Two figures frame the scene: a humorous dwarf and a bearded man of middle age whose teeth are bared in a grimace. Their attention is fixed upon a third figure, young, handsomely proportioned and serene, who sits between them. He is, it seems, a wild man, a man of the woods. Apart from a rude cloak he is naked, and his face is covered in hair – not just a beard, but locks that grow high on his cheeks and low on his forehead. The background foliage is lush, and a parrot, two monkeys and two dogs complete the bucolic scene. The whole thing could be an allegory of Nature were it not for the title, *Arrigo Peloso, Pietro Matto e Amon Nano* – Hairy Harry, Mad Peter and Tiny Amon – which tells us that it is really the inventory of a zoo.

The painting, commissioned by Cardinal Odoardo Farnese, was completed in 1599. It was only a trifle compared to the magnificent interiors of the Palazzo Farnese in Rome that the

HYPERTRICHOSIS LANUGINOSA. ARRIGO GONSALVUS,
ROME 1599. DETAIL FROM AGOSTINO CARRACCI, *ARRIGO
PELOSO, PIETRO MATTO E AMON NANO.*

Carraccis had already done for him. Attached to this palace,
which now houses the French Embassy, were a botanical garden
and a small menagerie, almost certainly the source of the
animals depicted in Agostino's painting. The wild man, a gift
from the Cardinal's kinsman Ranucci Farnese, lived there as
well. His cloak hints at his status and identity. It is a *tamarco*, the

robe of the Guanches, who once inhabited Tenerife in the Canary Islands but who had been briskly subjugated and largely exterminated by the Spanish a hundred years before.

Arrigo Gonsalvus, to give the wild man his full name, was not himself a Guanche. He was, however, the son of one, and a rather unusual one at that. In 1556 Petrus Gonsalvus arrived at the court of Henri II of France, brought there possibly as a slave from Tenerife. He could not have been more than twelve, but already a thick pelt of facial hair obscured his features. He seems to have been treated kindly there and was even given some education. In 1559, after the King's death, Gonsalvus appears at the court of Margaret, Duchess of Parma, despot of the Spanish Netherlands, where he married a young and rather pretty Dutchwoman who bore him at least four children of whom three were exceptionally hairy as well, among them Arrigo.

Margaret of Parma returned to Italy in 1582, the hairy family trailing in her wake. They were wonders, marvels of nature, and the Hapsburgs and Farneses could not get enough of them. Frederick II, Archduke of Tyrol, commissioned a set of individual portraits for his *Wunderkammer* at Schloss Ambras near Innsbruck where they may still be seen, part of his collection of natural curiosities. A group portrait of the family by Georg Hoefnagel appears in the illuminated *Bestiary* of Rudolf II, Emperor of Austria and Frederick's nephew, the only humans to do so. Perhaps the loveliest of the many portraits that depict this remarkable family is by the Bolognese painter Lavinia Fontana. It is of Arrigo's younger sister, Tognina, and shows the little hairy girl dressed in silvery brocades, smiling sweetly as she holds a document recounting her history aloft, and looking much like a preternaturally intelligent, if amiable, cat.

It may be thought that these portraits exaggerate the family's hairiness, but this is certainly not so. The travels of the family Gonsalvus in northern Italy were noted by that assiduous

HYPERTRICHOSIS LANUGINOSA. PETRUS GONSALVUS, AUSTRIA C.1582. UNKNOWN PAINTER, GERMAN SCHOOL.

encyclopaedist of nature Ulisse Aldrovandi, by then Professor of Natural History at the Papal University of Bologna. In his *Monstrorum historia*, he records meeting the family, describes them with care, and includes four woodcut portraits of them. Some scholars have suggested that Mad Peter, who stares so fixedly at the hairy man in Agostino Carracci's painting, is a portrait of Aldrovandi himself. In support of this charming conceit, it is certainly true that the bearded figure resembles Aldrovandi, and artist and naturalist had known each other since their student days. But in 1599 Aldrovandi would have been in his seventies, whereas Mad Peter is clearly in his vigorous prime.

Aldrovandi refers to Petrus Gonsalvus as the 'man of the woods' from the Canaries, and evidently believes that there were others like the Gonsalvus family there, a race of hairy people. There were not, of course. Petrus Gonsalvus was merely a man who happened to have been born with a mutation that caused a layer of hair to grow over parts of his face and body where in most people it does not. Nothing is known about the ultimate fate of Petrus, his wife, or his son. We do know that Petrus' daughter Tognina eventually married and bore several children who were as hairy as she.

Petrus Gonsalvus and his family were not the only hirsute people to have attracted royal curiosity. In 1826 John Crawfurd, British diplomat and naturalist, visited the Burmese capital of Ava to the north of Mandalay. On the throne was Bagydaw, scion of the Konbaungs, a family noted chiefly for the savagery of its dynastic struggles. (One of Bagydaw's predecessors had celebrated his succession to the throne of Ava in 1782 by slaughtering his brothers, their families, and some hundreds of his subjects – most of whom he immolated on a single gigantic pyre.) The Kongbaungs were also expansionist, a policy that attracted the ire of the dominant regional power, the British government in India. After the First Anglo–Burmese war, a humiliating peace was imposed upon the Burmese. The treaty was carried to Ava by Crawfurd, who found in Bagydaw's court a scene of medieval splendour complete with white elephants and human albinos. He also found Shwe-Maong.

'We had heard much,' wrote Crawfurd,

> of a person said to be covered all over with hair, and who, it was insisted upon more resembled an ape than a human being; a description, however, which I am glad to say was by no means realised in his appearance...The whole forehead, the cheeks, the eyelids, the nose, including a portion of the inside, the chin – in short, the whole face, with the exception of the red portion of the lips, were covered with fine hair. On the forehead and cheeks this was about eight inches long; and on the nose and chin it was about four inches. In colour it was of a silvery grey; its texture was silky, lank, and straight. The posterior and interior surfaces of the ears, with the inside of the external ear, were completely covered with hair of the same description as that on the face, and about eight inches long: it was this chiefly which contributed to give his whole appearance at first sight an unnatural and almost inhuman aspect.

Shwe-Maong was a Lao, a hills-man who as a five-year-old had been sent as tribute to Bagydaw's court by a local chieftain. Slightly built with mild brown eyes, he lived precariously, weaving baskets and playing the buffoon; as a boy he had been taught to imitate the monkeys that lived in the teak forests of the Burmese hinterland. When Shwe-Maong was in his early twenties, Bagydaw gave him a court beauty in marriage by whom he fathered four children, one of whom, a 'stout and very fine' girl named Maphoon, was also hairy. Born with hairy ears, by the time she was six months old the rest of her body was covered in fine grey down. When Crawfurd saw her she was two or three years old and her face was no longer visible. Thirty years after Crawfurd's account Maphoon appears again in the record of another diplomatic mission to Ava sent to deal with the ever-fractious Kongbaungs. By then she was a mature woman who looked much like her father, long since dead. Silky hair flowed over her face, leaving only her eyes and lips exposed; her neck, breasts and arms were covered with a fine down, and she also had her father's gentle manners. She had married – Bagydaw's successor, perhaps out of intellectual curiosity, had offered a reward to any man who would have her – and was the mother of two boys, both of whom were hairy as well. One of them later married, and a photograph that dates from perhaps 1875 shows three generations of the family – Maphoon, her son, and his daughter – all identically hairy.

In 1885 the British finally conquered Upper Burma in the Third Anglo–Burmese War, and the palace at Ava was destroyed. Maphoon and her family fled into the forests where, some weeks later, they were found by an Italian army officer who persuaded them to travel to Europe. And it is there, in the summer of 1886, that we last hear of Shwe-Maong's family, exhibiting themselves at the Egyptian Hall in Piccadilly and in Paris at the Folies Bergère.

THE TOPOGRAPHY OF HAIR

We are born with about five million hair follicles, and that is all we will ever have. The hair follicles are arranged in rows, adjacent follicles intercalated between each other in strict order. How does this regularity come about? If hair follicles were simply scattered randomly upon our scalps, each of us would have at least a few gaps in the thatch. The problem of how follicles come to be arranged with such precision is deep and difficult. It is the problem of how to make a regular pattern out of nothing.

The difficulty lies in the word 'regular'. It is fairly easy to imagine how an organism can make unique parts – five different fingers, for example. It is merely a matter of having pre-programmed cells respond to a single gradient in the concentration of some molecule. Our fingers are, indeed, specified in just this way. But what if, instead of a hand with five unique fingers, one wished to make a hand with only two alternating finger-types, say, ring fingers and index fingers? A strange variety of hand that looked something like this: ring-index-ring-index-ring? No such hand has ever existed. But this, in essence, is the problem that our skins present. Out of bland embryonic

Hypertrichosis lanuginosa. Maphoon,
Burma c.1856.

uniformity the skin must somehow order itself into a lattice of regularly spaced hair follicles separated by bits of skin. Clearly, some subtle device is needed.

The exact form of that device is still quite obscure, but the logic of its workings is not. What is needed is a way of making hair follicles, but of not making them everywhere. A foetus begins to develop the first of its follicles around three months after conception. Five million hair follicles do not appear all at once: instead, they début on our brows, then spread like a rash, first to the rest of the head and face, then down the neck, throat and torso, across the hips and shoulders, and finally down arms and legs.

I like the simile of a rash, for it suggests the spread of some infectious change in the skin cells, a change that expands outwards from a small beginning. This change transforms the cells of the skin from a quiescent state to one capable of producing follicles. It probably happens cell by cell. Perhaps it begins with just one cell somewhere on the forehead which induces the same change in its neighbours, which then transform their neighbours, and so on and so on. No one knows what the nature of the change is, but it is possible to make some guesses.

Each hair follicle is a chimera, a hybrid, of two different tissues. So is skin itself. The skin that we see, that we touch, and that weathers the elements, is the epidermis, a stratified layer of cells that originate in the outermost germ layer of the embryo, the ectoderm. Underneath the epidermis is another, thicker, layer, the dermis, which comes from the mesoderm. Dermis and epidermis are intimate collaborators in the making of a hair follicle. Their relationship is of the nature of talkers holding a conversation, a molecular dialogue of signal and counter-signal.

There is a simple, if slightly eccentric, experiment that shows this. In 1999, trading on a shared devotion to each other and to science, a married pair of scientists used each other as guinea pigs. They excised a piece of dermis from his scalp and then transplanted it to the hairless region underneath her arm. It may seem surprising that she didn't reject (in the immunological sense) her spouse's tissue, but it appears that hair follicles are somehow protected from immune-system surveillance. In the event, shortly after the wound healed she started growing long scalp hairs in the area that had received the transplant. The experiment showed that the dermis has a voice, one that tells the epidermis: 'make follicles here'. Indeed, the change that spreads like a rash across the foetus as it develops hair follicles is the dermal cells acquiring that voice in succession – a volubility that spreads to dermal cells everywhere bar those in the fingertips, palms, soles, lips and genitals, which for some reason remain silent.

If, in the conversation of the skin, the dermis's instructions are the opening gambit, it is one to which the epidermis has immediate right of reply. As dermal cells spring to life, urging the epidermis to make follicles, it must, with regularity and firmness, reply 'no'. Were it not to do so, the foetus's skin would become a single giant hair follicle, or perhaps a tumorous mass of malformed follicles and hairs. The way in which the epidermis counters the dermis is what gives hair follicles their precise spacing. Each newly formed hair follicle issues instructions that prevent the epidermal cells around it from also becoming hair follicles. Not only does each newly formed follicle prevent surrounding cells from hearing the dermis's insistent demands, it probably shuts them off at source.

The words in this conversation seem to be signalling molecules of the sort that we have come across before. Bone morphogenetic proteins are good candidates for the epidermal inhibitor. Bird feathers are distantly homologous to mammal hair, and if a bead soaked with BMP is placed on a chicken embryo's skin, the infected patch will not form feathers. If the same experiment is done with fibroblast growth factor, extra (albeit weirdly distorted) feathers will form – perhaps it is the original follicle-inducing signal. These molecules are thought to work in the same way in our hair follicles. But the signals around the developing follicle are so various, abundant and dynamic that it is difficult to know what they all do. We do know that mice engineered with defective hair-follicle signals are often bald.

GRASSLESS FIELDS

The one thing that many of us would dearly like to know about hair is why we lose it. Just how many men suffer from 'androgenetic alopecia' or 'male pattern balding' is a matter of definition, but claims that it can be detected in 20 per cent of American men in their twenties, 50 per cent of thirty-to-fifty-year-olds, and 80 per cent of seventy-to-eighty-year-olds seem about right. Balding is truly a white man's burden: Africans, East Asians and Amerindians (Native Americans) all have lifetime probabilities of balding lower than 25 per cent. Medically innocuous, it is a dispiriting disorder. When Ovid wrote in *Ars amatoria*: 'A field without grass is an eyesore/so is a tree without leaves/so is a head without hair,' he spoke for legions. For at least a century Americans have shown a marked aversion to electing bald men to their nation's highest office. Excluding Gerald Ford (1974–77), who was bald but not elected, the last bald president was Dwight D. Eisenhower (1953–61). Europeans have been more sympathetic to the bare-headed politico (Churchill, Papandreou, Simitis, Giscard d'Estaing, Mitterrand, Chirac, Craxi, Mussolini), but even they lagged behind the Soviets, who inexplicably installed, if not exactly elected, bald and hirsute leaders in strict alternation: Lenin (bald), Stalin (hairy), Khrushchev (bald), Brezhnev (hairy), Andropov (bald), Chernenko (hairy), Gorbachev (bald) – a tradition that has been maintained in the Russian Republic with Yeltsin (hairy) and Putin (comb-over).

What causes balding? Samuel Johnson's views on the matter – 'The cause of baldness in man is dryness of the brain, and its shrinking from the skull' – may be safely discounted, as can the theory, popular around 1900, that it was due to the wearing of hats. But dermatologists are hard pressed to offer more convincing explanations. Baldness obviously runs in families, but claims that it is due to a single recessive mutation or else 'inherited from the mother's side' (recessive X-linked) are wrong. Male pattern balding is caused by several genes, none of which has been yet identified. Whatever they are, they must affect the life-cycle of the hair follicle.

Hair follicles have the peculiar habit of periodically destroying and then reconstructing themselves. Most of the time they simply produce hair. A single scalp follicle can work on lengthening a hair for anywhere between two and eight years; the longer it does so, the longer the hair becomes. Mouse follicles work on a given hair for only two weeks, which explains why their fur is so short. When the follicle comes to the end of its growth period it begins to retreat within the skin and die, and the hair falls out. Halfway down the follicle, however, there is a bulge of epidermal cells – 'stem cells' – that have two remarkable properties: they are immortal, and they can become all the other types of epidermal cells of which the follicle is made. They are the stuff from which the follicle rebuilds itself.

But not in bald men. Instead of rejuvenating into a fully productive follicle, all that is produced is a pale and feeble imitation of the real thing; a follicular epigone capable only of making tiny hairs. Why this happens remains a mystery. One fact is, however, known: to go bald you need testosterone, and plenty of it. In the passage of *Historia animalium* in which Aristotle tells us that eunuchs are tall, he also says that they do not go bald, an observation confirmed in 1913 by a study of the last of the Ottoman eunuchs. The first rigorous demonstration that testosterone, rather than any other testicular hormone such as estrogen, is the culprit came from a 1942 study by the American physician James Hamilton. Some of the fifty-four eunuchs he studied were born without testes; some had been castrated as boys out of medical necessity (inguinal hernias, for example). Hamilton does not reveal where he found the rest of his experimental subjects, but one of his later papers suggests that they were mentally retarded men who had been castrated as boys in

Kansas mental institutions, a legacy of eugenic programmes that ran in the United States until the 1960s (and even later elsewhere). Consistent with Aristotle's claim, none of the men who had been castrated before their late teens developed any sort of baldness, not even the relatively high foreheads that nearly all mature men have. This wasn't because they all happened to come from families with good hair – several had balding male relatives. Proof that the eunuchs' boyish hairlines were due to their lack of testosterone came when Hamilton gave them male hormone supplements and some of them began to lose their hair. When he stopped the treatment, it promptly grew back.

The need for balding men to have their testicles is the likely origin of the idea that prematurely bald men are unusually virile. It is a claim that has the ring of wistful propaganda about it. (Even Julius Caesar, it is said, rejoiced in the title 'the bald adulterer'.) To be sure, there is a sad irony in the fact that the very hormone that gives men their beards in puberty denudes their scalps a few years later, but there is no evidence that prematurely bald men either have more testosterone than their hairier contemporaries or father more children. On the other hand, it is probably a lack of testosterone that prevents women from going bald. Women who acquire, for whatever reason, abnormally high levels of testosterone not only grow beards but tend to go bald as their baldness genes, hitherto silent, manifest themselves.

Is there any hope for the bald? Contrary to the folklore of depilation, shaving does not make hair grow faster, thicker, or darker – so there's no point removing what little you have left except on aesthetic grounds. More usefully, at least one of the baldness therapies currently marketed, said to be quite effective, is an inhibitor of dihydroxytestosterone (DHT), the more potent version of testosterone. If this doesn't appeal (and only a few users suffer impotence as a side-effect), then other therapies may soon be available. The resting hair follicles of a young mouse can be made to produce hair if dosed with a virus expressing high levels of sonic hedgehog. The surplus sonic probably forces the proliferation of the stem cells in the bulge of the hair follicle; if it could do the same for the crippled follicles on bald scalps, then a cure for baldness would surely be at hand. But maybe the hair follicles of bald scalps cannot be rejuvenated; if so, it will be necessary to make new ones. This may well be possible. Mice that have been engineered to overproduce a special form of the protein β-catenin make entirely new hair follicles at an age when normal mice don't. Unfortunately, both sonic hedgehog and β-catenin are extremely potent molecules. Excess amounts of either tend to produce hair-follicle tumors – the product of all those extra stem cells. It may be easy to spur skin to make new hair; rather harder to tame it.

BENEATH THE NAKED APE

Four centuries and two continents apart, Petrus Gonsalvus and Shwe-Maong are startlingly alike. Were Petrus to discard his richly sombre robes with their scarlet facings and knot a lungyi about his waist, the two men could be brothers. Nineteenth-century scientists such as Carl von Siebold and Alexander

Brandt were, however, more impressed by the resemblance of the hairy men to orangutans. Influenced by the new *Darwinismus* they suggested that hairiness was atavistic. This may seem like a version, albeit dressed up in scientific terminology, of the ancient equation between hairiness and bestiality. But the scientists were careful to note that though their subjects may have looked like apes, they were in fact quite human.

One can still, occasionally, come across claims that surplus-hair mutations reveal the fur beneath the naked ape. But there is reason to think that the atavism hypothesis is wrong – at least as applied to these two families. Both the hairy Burmese and Canary Islanders are described as having exceptionally fine, silken hair. This does not really resemble the robust pelt that covers adult apes – nor even human scalp or pubic hair. And hairy as great apes are, they are less so than the hairiest humans. Petrus and Shwe-Maong had noses, cheeks and ears that were covered in hair – exactly where great apes have rather little.

Where, then, does the surplus hair come from? One possible source is the foetus. Around five months after conception every human foetus grows a dense coat of hair. This 'lanugo' hair is fine, silky, less than a centimetre long, and enigmatically fleeting. Just weeks after it has grown it is shed again. Were it not for the occasional child born with lingering remnants of lanugo (often on the ears), we would hardly know that it was ever there. It seems likely that the mutation that afflicted the hairy families caused this lanugo to be retained. Instead of switching over to the normal pattern of juvenile, and then adult, hair production, their hair follicles were arrested in foetal mode.

And not just their hair follicles. In his description of Shwe-Maong, John Crawfurd notes that the hairy Burmese man had only nine teeth: four incisors and one canine in the upper jaw, four incisors in the lower, and no molars in either. Shwe-Maong's daughter, Maphoon, had even fewer. Careful inquiries showed that they had not lost their missing teeth: they had never grown them. It was as if their teeth and hair had simply come to a halt somewhere around the sixth month of foetal development even as the rest of their bodies marched on.

Darwin himself knew of the Burmese hairy family. In *The descent of man and selection in relation to sex* (1859) he cites the bribe needed to secure Maphoon a husband as proof that hairiness in women is universally unattractive. Nowhere, however, does he suggest that hairiness is an atavism. He is, instead, interested in the connection between hair and teeth. A Mr Wedderburn had told him of a 'Hindoo' family in the Scinde – modern-day Pakistan – in which ten men from four generations were almost entirely toothless, but, far from being hairy, were rather bald – and had been so from birth. The bald, toothless Hindoos also lacked sweat glands; unable to perspire, they wilted in Hyderabad's heat.

Hair, teeth, sweat glands and (though Darwin does not mention them) breasts, organs seemingly so various in their purpose and plan, are intimately connected. They are all places where skin has swollen or cavitated to make something new. The simple tube that is a hair follicle, the robust anvil of dentine and enamel that is a tooth, and the bulging burden of ducts that is a breast, are all variations on a constructional theme. A genetic disorder – there are more than a hundred – that affects one of these organs will often affect another.

These organs do not merely share an origin in skin; they are also made in much the same way. Even as hair follicles are forming throughout the foetal epidermis, other epidermal cells are clumping and cavitating to form teeth or mammary glands. Like the hair follicle, each of these skin organs is a chimera: part ectoderm, part mesoderm.

The kinship between all these organs can be seen in the molecular signals that make them. The 'Hindoos' still live near Hyderabad, where, confusingly, they are known as 'Bhudas' but are in fact Muslim. By 1934, six generations of Bhudas had spread across eight families. Now there are many more. Their distinctive appearance means that they recognise each other as relations, but the name of their mutant forebear seems to be forgotten. Just as Darwin's correspondent said, they have neither sweat glands nor teeth (except for the occasional molar), but they do have at least a little scalp hair. They carry a mutation in a gene that encodes a protein called ectodysplasin, named for the disorder its absence causes: *Ecto*dermal *dysplas*ia. A mutation in the same gene may also explain the Mexican hairless dog. Alias *El perro pelon* or the *Xoloitzcuintle*, the dog is said to have been bred by Aztecs in the fourteenth century, possibly for meat but more likely as a kind of bed-warmer. It, too, is bald, toothless and has dry and crinkly skin for want of sebaceous glands.

An even deeper organ-kinship is evident in an odd variety of aquarium fish. Since at least the start of the Tokugawa Shogunate in the early seventeenth century, Japanese fanciers have bred the Medaka, *Oryzias latipes,* a small fish that normally lives in rice-paddies. A sort of poor man's Koi, they can be bought from the night-stalls in Japanese cities where, among the varieties for sale – albino, spotted, long-fin – there are mutants that have no scales. The Medaka's nudity, like the Bhudas', is caused by a mutation that disables ectodysplasin signalling.

The use of a single molecule in the making of human teeth, hair follicles and sweat glands is a legacy of the evolutionary history that these organs share. This history is evidently also shared – at various removes – with the feathers of birds and the scales of fish and reptiles. All these organs have evolved from some simple skin organ possessed by some ancient, long-extinct ancestor of the vertebrates. No one knows exactly what this organ was. The best guess is that it resembled the tooth-like scales that give shark skin its roughness.

The right signal can even bring about the unexpected resurrection of organs long buried by evolution. Birds don't have teeth, but their dinosaur ancestors certainly did. If a piece of ectoderm from a foetal chicken's beak is grafted onto a piece of mesoderm from a foetal mouse's mandible, and both are placed in the eye-orbit of a young mouse, the chicken tissue, which has not seen a tooth for sixty million years, suddenly begins to make them: hen's teeth, shaped something like tiny molars, complete with dentine and enamel. This implies that the molecular signals used by *Tyrannosaurus rex* to make its mighty fangs are the same that a mouse uses to make its miniature molars. Signals that chickens just seem to have lost.

*   *   *

Perhaps it is also the retrieval of an ancient signalling system, partly buried by evolution, that causes some people to have extra nipples or even breasts. Humans and great apes have only two nipples but most mammals have many more. Sometimes extra nipples are little more than a small dark bump somewhere on the abdomen; at other times they are fully developed breasts. They are common: between 2 and 10 per cent of the population have at least one. In Europeans extra nipples or breasts are usually found somewhere below the normal ones, often in a line running directly down the abdomen. Japanese women, curiously, seem to get them above the normal breasts, often in the armpits.

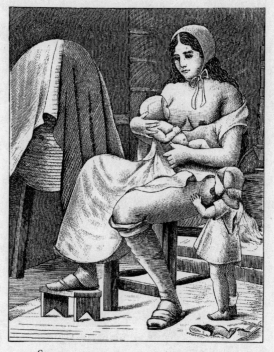

SUPERNUMERARY BREAST ON THIGH.

These patterns of extra nipples may recollect an ancient 'milk line' – a row of ten pairs of teats that ran from the armpits to the thighs in some ancestral mammal. Armpit breasts are found in the lemur, *Gaelopithecus volans*, and the record number of nipples found on a single person seems to be nine (five on one side, four on the other). Wherever they are, extra breasts often work like normal ones, swelling and even lactating during pregnancy, and there are even accounts of women suckling children from super-numerary thigh-breasts. Extra nipples and breasts run in fami-lies, though the mutation (or mutations) that causes them has not been identified. However, a group of London researchers are attempting to determine the mutation behind a strain of mice that have eight nipples instead of the usual six. They have already dubbed the gene *Scaramanga* – for the villain of the James Bond film *The Man with the Golden Gun* who had, as a mark of his depravity, a supernumerary nipple on his upper left chest.

ARTEMIS EPHESIA

Breasts bring us back to Linnaeus. In 1761, made famous by *Systema naturae*, Linnaeus published one of his lesser-known works, a synopsis of the Swedish animals called *Fauna svecica*. The name was revolutionary: it was the first time that the word 'fauna' – from the Roman name for Pan-the-God – had been used to describe a work of this sort; a direct counterpart to the 'floras' that were already proliferating. As a frontispiece to this work Linnaeus chose a curious emblem, a representation of the Greek goddess Artemis, or Diana, of Ephesus. We don't know why he picked this particular emblem, but there are several possibilities.

Artemis Ephesia was, in the inexplicably duplicitous way of Greek deities, goddess of both nature and cities. In her original incarnation as the object of a cult that flourished in Asia Minor from around the sixth century BC, her image was hung on city walls to protect them from evil, while being surrounded by icons

ARTEMIS EPHESIA IN SWEDEN. FRONTISPIECE OF LINNAEUS 1761 *Fauna svecica*.

of the country: garlands of vines and climbing animals such as lions, snakes, birds and harpies. Retrieved from the ruins of Ephesus, the eighteenth century made her into a symbol of wildness and of reason. The Jacobins even dedicated a Temple of Reason to her that once stood in Strasbourg, but is now gone. Perhaps this is why Linnaeus placed her at the front of his *Fauna* – as a symbol of the mastery of Reason over Nature, albeit a Swedish nature, in which, far from her Mediterranean home, Artemis stands among browsing reindeer.

But perhaps she had another, more direct, meaning for Linnaeus as well. What is most striking about his Artemis are not the animals that surround her, but her four prominent breasts. In this she is a direct echo of the statues of her in antiquity, all of which are laden with a varying number of thoracic and abdominal protuberances. In the Renaissance these bumps were invariably interpreted as a case of extreme polymastia, but more sceptical modern scholars say they are more likely to have simply been strings of dates, bulls' testicles, or perhaps just part of the cuirass in which the goddess was clad. Be that as it may, Linnaeus' Artemis obviously has four fine breasts, and it seems quite possible that they are a direct allusion to one of his finest inventions, the Mammalia. For Linnaeus made the presence of mammary glands one of the defining features of what we are: members of that great class of creatures that embraces simultaneously the pygmy shrew and the blue whale.

There is a third possible source of Linnaeus' Artemis, one that brings us back to where we started – the way in which we differentiate ourselves from the rest of brute creation. When describing a species, Linnaeus did what taxonomists still do – he listed the things that distinguish it from all others. For all species, that is, but one: our own. When it came to *Homo sapiens*, instead of speaking of the number and kinds of teeth we have, the density of our hair, the distribution of our nipples, Linnaeus wrote only this: *Nosce te ipsum*. In a footnote he says that these are the words of Solon written in letters of gold upon the temple of Diana. Perhaps in choosing Artemis Ephesia as his iconographic symbol, Linnaeus is remembering and alluding to this account of the human species, the most concise possible: know thyself.

That is where Linnaeus' discussion of *Homo sapiens* ends, but for a few strangely exigent epigrams in which he instructs us in the meaning of the new identity that he has given us. 'Know thyself,' he says, created by God; blessed with minds with which to worship Him; as the most perfect and wonderful of machines; as masters of the animals; as the lords of creation – all sentiments that today ring with the poignancy of certainties long since gone. Yet it is his parting shot that is most telling, and that could be taken as epigrammatic of much of what I have written here:

> *Know thyself, pathologically, what a fragile bubble you are, and exposed to a thousand calamities.*
>
> *If you understand these things, you are man, and a genus very distinct from all the others.*

IX

THE SOBER LIFE

[ON AGEING]

Huntington disease is one of the nastier neuro-degenerative syndromes. It usually first appears as a mild psychosis and does not seem especially serious. But, as the disease progresses, the psychotic episodes increase in frequency and severity. Motor-coordination also deteriorates, a characteristic rigidity of gait and movement sets in and then, eventually, paralysis. In the disorder's final phase, which can take up to ten or twenty years to appear, the patient becomes demented and experiences neural seizures, one of which is eventually fatal. The disease is caused by dominant mutations that disable a protein used in synaptic connections of the brain's neurons. For reasons that are not fully understood, the mutant form of the protein initiates a molecular programme that gradually kills the neurons instead.

LUIGI CORNARO (1464–1566). TINTORETTO.

Huntington disease has several strange features. One is the way in which its symptoms become more severe from one generation to the next. This phenomenon, called 'anticipation', arises from a peculiarity of the Huntington gene itself and the mutations that cause the disease. The gene contains a region in which three nucleotides, CAG, are repeated over and over again. Most people have between eight and thirty-six of these repeats. Huntington disease mutations increase the number of repeats, so disordering the structure of the protein. Several mutations of this sort cripple the protein ever further over successive generations, increasing the severity of the disease.

Another oddity of Huntington is its frequency. It afflicts about 1 in 10,000 Europeans. This is very high – most dominant mutations that kill have frequencies of about one in a million. But Huntington disease can persist in a family for generations. In 1872, George Huntington, a New York physician, described the disorder from families in Long Island, New York. Among their ancestors was one Jeffrey Ferris who emigrated from Leicester, England, in 1634. He almost certainly had the disease, as do many of his descendants today. In South Africa, about two hundred Huntington's patients are descended from Elsje Cloetens, the daughter of a Dutchman who arrived with Jan van Riebeeck to found the Cape Colony in 1652. A large group of Huntington's patients who live near Lake Maracaibo, Venezula, are the decendants of a German sailor who landed there in 1860.

How can so lethal a disorder transcend the span of so many generations? In 1941 the brilliant and eccentric British geneticist J.B.S. Haldane proposed an answer. He pointed out that, unlike most genetic disorders, the symptoms of Huntington disease usually appear in middle age. By this time most people with the defective gene have had their children — each of whom will have had a 50 per cent chance of inheriting the defective gene. Unlike most lethal dominant mutations that kill in childhood and so are never transmitted to the following generation, the Huntington mutation hardly impairs the reproductive success of those who bear it. Middle age is almost invisible to natural selection.

Few other disorders caused by a single mutation have such devastating effects so late in life. Yet the strangeness of Huntington disease is deceptive, for Haldane's explanation of why it is so common also explains, with a little generalisation, why we, and most other animals, age. In this chapter I will argue that ageing is a genetic disorder, or rather, it is many genetic disorders, some of which afflict us all, others of which afflict only some of us. This point of view goes against the grain of most definitions of disease. Medical tradition distinguishes between 'normal' ageing, about which nothing much is done, and 'age-related diseases', such as arteriosclerosis, cancer and osteoporosis, that consume vast amounts of national health budgets. But this distinction is an illusion, a necessary medical fiction that allows physicians to ignore a disease that affects us all but which they are impotent to cure or even ameliorate. Properly understood, ageing is precisely what it seems: a grim and universal affliction.

IMPOTENT SELECTION

Ageing is the intrinsic decline of our bodies. Its most obvious manifestation is the increased rate at which we die as we grow older. An eight-year-old child in a developed country has about a 1 in 5000 chance of not seeing her next birthday; for an eighty-year-old it is about 1 in 20. Of course, it is possible to be killed by causes quite unrelated to ageing – violence, contagious disease, accidents – but their collective toll is quite small. Were it not for ageing's pervasive effects, 95 per cent of us would celebrate our centenaries; half of us would better the biblical Patriarchs by centuries and live for more than a thousand years. We could see in the fourth millennium AD.

The evolutionary explanation for why we, and most other creatures, age rests upon two ideas, both implicit in Haldane's explanation for the frequency of Huntington disease. The first is that the ill-effects of some mutations are felt only late in life. Most obviously a mutation might cause a slow-progressing disease. The Huntington mutation is just such a time-bomb. So is the SOST mutation that causes sclerosteosis in Afrikaaners; children are relatively unaffected but the excess bone growth kills in middle age. So are mutations in BRCA1, the familial breast-cancer gene whose ill-effects are usually felt only by women in their thirties and forties. And so is a variant of the APOE gene called ε-4 that predisposes elderly people to heart attacks and Alzheimer's.

Such examples could be multiplied, yet it must be conceded that not a great deal is known about the time-bombs with the longest fuses, those that detonate past middle age and that cause senescence. For the moment, let us simply suppose that they exist. To do so, however, is not sufficient to explain ageing. It is also necessary to understand how it is that time-bomb mutations have come to be such an inescapable part of human life. Haldane alluded to the explanation for this when he argued that the Huntington mutation is not seen by natural selection. The same logic can be applied more generally. Imagine a dominant mutation that renders a twenty-year-old man impotent for the rest of his life. In twenty-first-century Britain at least, relatively few men have fathered children by the age of twenty, and after age twenty, the victim of such a mutation will never do so. Whatever he may accomplish in the course of the rest of his life, as far as genetic posterity is concerned he may as well never have been born. The same mutation may occur many times in many men but it will, adolescent fathers aside, never be transmitted to future generations and so will always remain rare. Imagine now another dominant mutation, one that also renders its carrier impotent, but does so only at the age of ninety. For such a man, the odds are excellent that he will be quite oblivious to his loss for the simple reason that he will be dead, having been previously claimed by cancer, a cardiac infarction, influenza, or a failure to notice the approach of the Clapham omnibus. Six feet under, the cost of Viagra is not an issue. Alive and virile he will, however, have sired any number of children, some of whom will bear the mutation, as will some of their children, and so on. Indeed, it is quite possible that the mutation will, simply by chance, spread throughout the population so that, after many generations, all men will be impotent at age ninety – essentially the case today.

This argument is just a restatement of Haldane's: that the force of natural selection against deleterious mutations declines over the course of life. But it was another British scientist, Sir Peter Medawar, who first generalised this to explain the diversity of ways in which our bodies break down while ageing. Late in life, some mutations impair our cardiovascular fitness, others our resistance to cancers or pathogens, others virility, yet others our wits. Such long-fuse mutations have afflicted us forever and, unimpeded by natural selection, they have spread and become universal.

Medawar's explanation of the ultimate causes of ageing surely has a great deal of truth to it, but it has one weakness, and that is its appeal to chance. It is easy to see why mutations that cause some grievous error in old age are not selected against, but is that absence of impediment enough to account for their spread throughout humanity? Perhaps. There are probably thousands of different mutations that have ill-effects late in life, and each of these must have occurred incalculably many times in human history. It is certainly plausible that some spread by chance, particularly at times when population sizes were small.

But an appeal to chance is never satisfying; we would prefer a deterministic theory. In 1957, an American evolutionary biologist, George Williams, proposed one. He argued that the mutations that cause ageing spread not by chance but because they confer some benefit, albeit only to the young. Imagine, once again, a mutation that causes impotence at age ninety, but that also confers unusual virility at age twenty. The carrier of such a mutation might well sire more children than other men, and so the gene would spread. In the calculus of natural selection, small benefits reaped early often outweigh severe costs paid later on. Old age, in this view, is the price we pay for the lavish beauty and exuberant excess of youth.

Some geneticists have used this logic to explain why Huntington disease is so common. They argue that women with the disorder are, in the first stages of their disease, unusually promiscuous, or feckless, or at least unusually fecund. One study has shown that women with Huntington disease have more illegitimate children than their unaffected siblings. Perhaps, the argument goes, the disorder causes unusually high levels of gonadotropin, a hormone that influences sexual behaviour. There is little evidence to support any of this.

More generally, so little is known about the genes that cause human ageing that it is difficult to know whether Medawar's or Williams's view is the more accurate. In a way, the difference between the two theories does not matter; they may both be right, for they are similar in their causes and their consequences. Both propose that ageing is not *for* anything, but is, instead, just an epiphenomenon of evolution. It is ultimately due to the inability of natural selection to act against the mutations that cause disease in the old. Neither theory says much about the mechanical or molecular causes of ageing. They do not point to any one molecular device that we can fix and so ensure our immortality. Rather, both suggest that no such device will be found, and imply that ageing is the collective consequence of many different mutations that gradually wear down and then destroy our bodies.

Perhaps this is why, despite much effort, the mechanistic causes of ageing remain so elusive. The root of ageing's evil has been claimed, at one time or another, to lie in any one of a dozen aspects of human biology. Some have claimed that it is caused by the fermentation of bacteria in our guts; others by a slow-down in the rate at which the body's cells divide; yet others have pointed to the exhausting effects of bearing and raising children. Others again have proposed that ageing is caused by the exhaustion of some vital substance, or else that chemicals produced by our own cells gradually poison us. Many of these ideas are probably absurd, but some probably contain at least an element of truth. What follows is a survey of some of the most plausible ones: a brief history of decay.

GERONTOCRATS

In his declining years, flush with cash and fame from having invented the telephone, Alexander Graham Bell turned his attention to genetics. His first efforts were modest. He bred a variety of sheep with four nipples instead of the usual two. Then, combining his interests in sound and heredity, he studied the genetics of deafness. But his passion was the genetics of human longevity. He began with the family of one of America's Pilgrim Fathers, a William Hyde (settled Norwich, Connecticut, in 1660), whose descendants, all 8797 of them, had been traced by genealogists. Analysing their records, Bell concluded that longevity was mostly inherited. Neither his data nor his statistics justified this conclusion. But he wasn't far wrong – modern estimates put the heritability of European longevity between 20 and 50 per cent. In the event, it was enough to set him off on far grander plans.

Like many early-twentieth-century scientific men, Bell was an enthusiast of eugenics. Not 'negative' eugenics – the state-enforced sterilisation of the mentally disabled and the antisocial – that were vogueish in the 1920s, for this he found repugnant. Bell was a humane man; it is not for nothing that America's premier organisation for the deaf bears his name. His view of eugenics was more 'positive', liberal, indeed entrepreneurial: he saw it as an instrument in the marketplace of human affections. Bell proposed, and then began, the compilation of vast numbers of longevity records from Washington, DC, area schools. His idea was to ask children how old their parents and grandparents were, and then publish the results along with their names and addresses in a volume that he called, without equivocation, a 'human stud-book'. People, he thought, would be sure to consult his stud-book; the descendants of long-lived individuals would search each other out, fall in love, and breed. What of the descendants of short-lived people? Perhaps they would simply remain unmarried. Or perhaps long-lived and short-lived people would separate into distinct races; there would be true gerontocracy. Genetic progress, like economic progress, requires efficient markets, and efficient markets need information; it was all very clear.

* * *

Alexander Graham Bell's scheme was visionary and only slightly mad. (Who among us would choose the object of our desires on the basis of mean grandparental longevity?) Unsurprisingly, it foundered with his death in 1922. Yet had the scheme become universal, and had people behaved as Bell hoped they would, the results would surely have been spectacular. There is no doubt that the careful breeding of long-lived families would, with time, have resulted in a strain of long-lived people. Perhaps not patriarchially long-lived, but a good deal longer than the seventy-something years that is all we can reasonably hope for. We can guess this, because experimental schemes, not too different from Bell's, work in other creatures.

In the 1980s the evolutionary account of ageing given by Williams and Medawar inspired researchers to attempt the creation of a breed of long-lived fruit flies. If the ultimate cause of ageing lay in the absence of natural selection late in life, they reasoned, perhaps long-lived flies could be produced by forcing natural selection upon old flies. A fruit fly can breed at two weeks of age, almost as soon as it emerges from its pupa, but by ten weeks it is quite old, perhaps as old as an octogenarian human. Male fruit flies never survive to this age, and the few females that do, the hardy survivors, have depleted metabolic reserves, tattered wings and feeble legs.

They can, however, lay at least a few eggs. And so populations of fruit flies were bred, generation after generation, only from the eggs of the oldest flies. The effect of this was to favour genetic polymorphisms that promoted survival and fertility at old age. As these increased in frequency, the flies evolved ever-longer lifespans. The speed at which this happened was remarkable. Ten generations of selective breeding were enough to increase the average longevity by 30 per cent – in human terms the equivalent of raising life expectancy from seventy-eight to just over a hundred. Fifty generations of selection, and life expectancy doubled.

Closer examination of these long-lived fruit flies showed that they were amazingly hardy. Deprived of food or water or subjected to noxious chemicals, they survived where shorter-lived flies expired. But glory in old age exacted a cost. As the flies' longevity evolved ever upwards, fertility in early life declined. Females laid fewer eggs, males were less inclined to mate. Eschewing profligacy, long-lived fruit flies hoarded their resources and established reserves of fats and sugars instead. They became sluggards, moving, breathing and metabolising slower than normal flies.

This result was just as predicted by George Williams's theory. If ageing is the genetic price of early-life reproductive success, then, conversely, increased longevity must be bought at the cost of a vigorous and fertile youth. This implies a simple economic relationship between fertility and longevity. A fly has only so many resources; it may use them to live to an old age or it may expend them on its progeny, but it cannot do both. It's a line of argument that goes back to Aristotle. In his account of animal physiology he supposed that animals need 'moisture' to live, and that they had a limited amount of it: life is warm and wet, and death is cold and dry. 'This is why,' he writes, 'animals that copulate frequently and those abounding in seed age quickly; the seed is a residue, and further, by being lost, it produces dryness.'

Since Aristotle, numerous studies have confirmed that reproduction exacts survival costs in a variety of creatures. The severity of these costs at the limit is shown by *Antechinus stuarti*, an Australian marsupial mouse. For the males of these mice, existence is little more than sex. Their brief adult lives consist of fighting other males, wandering about in search of females and, when they find them, engaging in exhausting twelve-hour-long copulations repeated daily for nearly two weeks. Perhaps unsurprisingly, after a single mating season they are dead, their tissues showing all the signs of catastrophic senescence. By the time they are done, they are devoid of sperm, their prostate glands have shrivelled up, their testes have become invaded by connective tissue, their adrenal glands are hypertrophied, their livers necrotic, their gastric tracts are haemorrhaging, and their penises are quite flaccid.

Marsupial mice are an especially blatant illustration of the idea that ageing is the consequence of youth's excesses. But there is evidence that the same economic principle affects humans, albeit to a more modest degree. The British have, of course, no Pilgrim Fathers to genealogise. Instead they have an aristocracy, mostly dating from Norman times, whose singular, indeed defining, virtue is an obsession with their own line of descent. Traditionally, the genealogies of Britain's noble houses have been recorded in the volumes of *Burke's peerage*, but these days a handier account of the pedigrees of most British peers, from the Dukes and Earls of Abercorn to the Barons of Willoughby de Broke, is available on CD-ROM. This database, which stretches back to 740 AD, contains, in so far as they are known, the birth dates, marriages, and progeny of the British nobility, and has been used to test the idea, evident to the parents of any newly born infant, that having children takes years off your life.

Before the Industrial Revolution, the wife of a British peer could expect to live to the age of forty-five. She could also expect to bear two or three children. These averages, however, conceal much variety in the chances of life. Some women died young, and so had very few children. Some died in the decade or two after menopause (fifty to sixty): on average they had 2.4 children. But some – albeit rather few – survived past age ninety. These elderly women had had, on average, only 1.8 children, and nearly half of them were childless.

This is a fascinating result. Not only is it consistent with the results of the fruit fly experiments, it suggests that had Alexander Graham Bell's dreams ever come to fruition, his gerontocrats would have had an ever dwindling fertility. A more sobering thought is that many, though surely not all, aspects of the senescent decline of our later years may be difficult to meliorate without damping down the physiological and sexual excesses of youth. In the future, humans may well be able to engineer themselves, be it by better drugs or better genes, to live as long as they please, but the cost may be twenty-year-olds with all the vigour, appetites and charm of the middle aged.

LA VITA SOBRIA

Is there a recipe for long life? Luigi Cornaro thought there was. In 1550, the Venetian nobleman published a tract called

Discorsi della vita sobria (Discourses on the sober life) in which he outlined the regime that had ensured his own longevity. He was probably eighty-three at the time, and lived until ninety-eight or 103 – there is some dispute about his birth date, though all agree that he reached a great age. By his own account he had, until the age of forty, lived a life of sensual dissipation. The consequences were pains in the stomach and side, gout, fever, and an unquenchable thirst. His physicians warned him that he must reform or die. He took their advice to heart and thenceforth devoted himself to a temperate and orderly way of life.

The chief ingredient of his new regime was simple: eating less, and then only what he found agreeable. 'Not to satiate oneself with food is the science of health,' he wrote. He is vague on specifics, but at the one point at which he reveals what his actual diet was, it does not sound too arduous. A typical meal would begin with bread, then a light broth, perhaps with an egg. But, he said, 'I also eat veal, kid and mutton; I eat fowls of all kinds, as well as partridges and birds like the thrush. I also partake of salt-water fish as the goldney and the like; and, among the ovarious fresh-water kinds, the pike and others.' A modest diet by sixteenth-century Italian standards then. Yet at one point he grew so thin that his friends urged him to eat more. Cornaro's oracular reply was that whosoever wished to eat long must eat little.

This is a little smug, but the *Vita sobria* charms – Cornaro is so clearly delighted by his longevity. A portrait by Tintoretto shows him in his splendid dotage, a grave and fine-featured patrician with skin made translucent by age. Cornaro spent his last years at his Paduan palazzo with its decorations by Raphael and at his villa in the Euganean Hills by the River Brenta with its exquisite gardens and fountains. 'I did not know,' he writes, 'that the world could be so beautiful until I was old.'

The *Vita sobria* was a huge success. As he grew older, Cornaro added material to its successive editions: two, three, and finally four *discorsi*. A product of the Italian Renaissance, the book's style was classical (Jacob Burckhardt cited it for its perfection), its physiology Aristotelian (much about moisture loss), and its sentiments Ciceronian (old age is a thing to be welcomed, a time of wisdom when passions have been burnt away). Its influence was long-lasting and can be found, for example, in the writings of the German physician Christian Hufeland, whose *Makrobiotik* (1796) outlines the theory from which every modern health-food fad ultimately derives.

The worst of it is that there is an element of truth in Cornaro's claim that the route to great longevity is eating less. By this I do not simply mean the sort of diet that will stave off gross obesity or even middle-age spread, but serious dieting of a sort that few people could sustain voluntarily. The only reliable way to extend the general physiological life of a mammal is to give it no more than two thirds of the daily calories that it wants. Dozens of studies have shown that 'caloric-restricted' mice live anywhere between 10 and 50 per cent longer than those which are allowed to eat as much as they want. Age for age, they are friskier, glossier and healthier than their controls. And they are slimmer: about half the weight of controls. Caloric-restricted mice do, of course, eventually die, but the ages at which they get diabetes, infections, renal malfunctions, autoimmune attacks, musculoskeletal degeneration, cardiomyopathy, neural degeneration and, most amazing of all, cancer are all delayed. Studies on rhesus monkeys are now under way to see if caloric restriction extends life in primates, but it will be another decade before we know the answer.

Uncertainty has not stopped many neo-Cornarists from committing themselves to lives of rigorous dieting. Caloric restriction has become a health fad like any other, with its own books and gurus. The diet usually consists of about a thousand calories per day, which is necessarily supplemented with a battery of vitamins and minerals. A thousand calories is about the minimum number needed to sustain the life of an average-sized man, though not enough to sustain his sex drive (or, to judge by pictures, his sex appeal). Whether these ultra-puritans will reap their reward is an open question. The severe caloric restriction experienced by the Dutch population during the *Hongerwinter* of 1944–45 certainly had no detectable beneficial effect on the long-term mortality rates of the survivors, but it could be argued that it takes decades of near-starvation for its virtues to become apparent.

Caloric restriction works in rats, mice, fruit flies and nematode worms. Why it does so remains mysterious. One explanation goes back to the deleterious effects of reproduction. Caloric-restricted animals have fewer offspring than those allowed to eat all they want; perhaps the energy savings that come with not reproducing are enough to ensure longevity. But there is probably more to it than this. In caloric-restricted fruit flies not only are the genes involved in reproduction largely switched off, but those involved in resistance to infection (the fly's immune system) are turned on, so that immunity proteins are produced at higher levels than they would be normally. This result suggests at least two reasons for the longevity of caloric-restricted animals. There may be many others besides. About two thousand of the fifteen thousand genes in the fly's genome show a response to caloric restriction. It is quite possible that caloric restriction works its magic by the cumulative benefits of dozens of different molecular pathways.

This should hardly come as a surprise. Evolutionary theory predicts that ageing is caused by the independent destruction of many different systems; if caloric restriction has such pervasive effects on health, then it too must work by maintaining the body in many different ways. Even so, many gerontologists still seek a single explanation for all the diverse manifestations of ageing and the way in which caloric restriction delays them. One idea is that ageing is caused by a kind of insidious poison that is a consequence of the very condition of being alive.

THE BREATH OF DEATH

'We term sleep a death and yet it is waking that kills us,' observed Thomas Browne in his *Religio medici*. That living itself is the cause of our decline – either by exhausting some vital substance or else by gradual self-poisoning – is one of the oldest ideas in the history of ageing science. In its most recent version, ageing is caused by small, pernicious molecules capable of oxidising DNA, proteins, lipids, indeed almost anything they come into contact with. In the course of normal respiration, oxygen is reduced to water. But this is an imperfect process, and several other molecular species called 'free radicals' are produced as by-products. These molecules, which have chemical formulas such as •OH (the • signifying an unpaired electron), are especially abundant in mitochondria, the sub-cellular structures in which respiration takes place. From there they leak into the rest of the cell, attacking other structures as they go.

The free radical theory postulates that ageing is caused by the accumulated damage that these molecules inflict upon cells over the course of years. An abundance of correlative evidence supports this. Free radicals certainly damage cells, and the kind of damage they do becomes more common in old age. Most disturbingly, they cause mutations. The DNA of each human cell receives ten thousand oxidative hits per day. While many of these are repaired, many are not. Old rats have about two million mutations per cell, about twice as many as young rats do. Most of these mutations will have no effect on the health of a given cell. But should the radical hit a gene vital for the survival of a cell it might well kill it. Should it hit a proliferation-control gene in a stem cell it might initiate a cancer. Should it hit a gene in the cells that give rise to rise to sperm and eggs, it may be transmitted to future generations.

Free radicals are clearly pernicious. But do they cause some or all of ageing? Perhaps. Long-lived animals – be they innately so or else calorie-restricted – seem to be exceptionally resistant to toxins such as paraquat, a weed-killer that works by inducing the production of free radicals. More direct evidence comes from genetic manipulations in a variety of animals. Animal cells contain a battery of defences against free radicals, among them a group of enzymes devoted to scavenging free radicals, the superoxide dismutases. Several different kinds of evidence suggest that they protect against some of ageing's effects.

An especially active form of superoxide dismutase seems to contribute to the longevities of the fruit flies, alluded to previously, that were the result of generations of gerontocratic reproduction. The founding population of these flies was polymorphic for two varieties of superoxide dismutase. Selection changed the frequencies of these variants so that the more active form became much more common in the populations of long-lived flies than in the short-lived ones. This wasn't just a matter of chance: the experiment was replicated five times, and the same result was found each time. In an even more direct demonstration of the benefits of this enzyme, flies were engineered to express human superoxide dismutase – apparently more potent than the fly's own – in their motor neurons. They lived 40 per cent longer than un-engineered controls, a particularly interesting result for it implies that superoxide dismutase can protect the nervous system. Finally, in the last few years many mutants have been found in nematode worms and fruit flies that seem to confer extraordinary longevity (one of them has even been named *Methuselah* after the patriarch who, Genesis assures us, lived to the age of 969). These mutants do not alter the sequences of superoxide dismutase genes themselves but rather affect genes that control when and how superoxide dismutase is activated. It is, it seems, hard to make a long-lived fly or worm without boosting superoxide dismutase by one means or another.

All these results suggest the following chain of argument: extra superoxide dismutase postpones ageing (at least in worms and flies); superoxide dismutase protects against free radicals; hence free radicals cause ageing. Does this imply that the means for postponing ageing in humans are at hand? Might we not simply engineer ourselves with a more effective superoxide dismutase and so gain years of life? The short answer seems to be no. Moreover, the reason that this won't work casts some doubt upon one of the premises of the foregoing argument.

Our genomes contain three genes that encode superoxide dismutases. Mutations in one of these, SOD1, have been known for years. These mutations are gain-of-function and dominant: they give a hyperactive protein. It may be thought that this is precisely the sort of mutation that, by analogy with fruit flies and worms, might give a human lifespan of 120 years. In fact, they kill by the age of fifty or so. SOD1 mutations cause amytrophic lateral sclerosis (ALS), a particularly ferocious neurological disease in which the motor neurons of the spinal cord, brain stem and motor cortex are progressively destroyed, leading to paralysis and death. In America the disorder is known as Lou Gehrig disease after the baseball player who suffered and died from it. Nowhere is the issue of physician-assisted suicide as pressing as it is in ALS.

These mutations pose a paradox. They suggest that superoxide dismutase kills motor neurons in humans, even as it protects them in flies. Why? For the last ten years this paradox has been resolved along the following lines. Superoxide dismutase is only the first step in an enzymatic pathway that neutralises free radicals. It converts the free radical oxygen anion, $O•_2$, to another molecule, H_2O_2, more commonly known as hydrogen peroxide, whose destructive effects upon biological tissue can be gauged by its fame as the active ingredient in chemical drain-cleaners and the classic suicide blonde. It takes another enzyme, catalase, to neutralise hydrogen peroxide by converting it to water. Perhaps an imbalance in the activity of these two enzymes in humans, but not flies, leads to a build-up of hydrogen peroxide in neurons and kills them.

It is a reasonable explanation, but it appears to be quite wrong. The reason that SOD1 mutations kill motor neurons has nothing to do with free radicals or hydrogen peroxide poisoning. Rather, their deleterious effects seem to be related to some other, slightly mysterious, role that superoxide dismutase has in the brain. Neurons are strange cells. They are large, have long protrusions called axons, and a whole special cellular architecture that goes with this. Besides scavenging free radicals, superoxide dismutase appears to have some role in this architecture. Biologists have adopted a lovely phrase to describe such multi-tasking proteins – they call them 'moonlighters'. Moonlighting SOD1 may also contribute to another neurological disorder, Down's syndrome. Children with Down's syndrome have three copies of chromosome 21 – the chromosome on which the SOD1 gene resides – instead of the usual two. Hundreds of different genes reside on this chromosome, and any or all of them might contribute to the distinctive features of Down's (mental retardation, the facial abnormalities, heart problems to name but a few), but the extra copy of SOD1 has long been fingered as one of the more destructive.

If superoxide dismutase moonlights, then the argument proposed above is predicated on a false premise. And with it goes one of the few good reasons for believing the whole free radical theory of ageing. The proponents of this theory (and among scientists they surely number in the thousands) may well feel that this is a harsh assessment of the only mechanistic account of the origin of ageing that has any pretensions to generality. It is certainly still possible that superoxide dismutase's seemingly beneficial effects on ageing are mostly due to free radical scavenging, but this remains to be shown. For the time being, however, few would disagree that superoxide dismutase can be struck from the list of elixirs that might one day stave off the decline of our later years.

A WRINKLE

Even if free radicals are not the sole, or even major, source of mutations, mutations may still cause at least some aspects of ageing. Mutations may be especially destructive in those tissues, such as skin, whose cells divide continually throughout life. Some of us keep relatively youthful complexions well into old age, while others wrinkle when young. This variety partly depends on the exposure to the elements, sun most obviously, that each of us has received; ultraviolet light is a powerful mutagen. But even sheltered skin ages. And for all the parasols, veils and sun-block in the world, no thirty-five-year-old's skin has ever glowed as it glowed when she was fifteen.

Wrinkling is a manifestation of a deeper inability of epidermal cells to replace themselves and maintain the integrity of the connective tissue of our skins. It is a problem that pervades our bodies. This is evident from people whose skins and connective tissues age with unusual, indeed catastrophic, rapidity. An inherited disorder called Werner's syndrome causes its victims to go grey and bald when still in their teens. In their twenties, the testicles atrophy in men as the ovarian follicles do in women – a kind of premature menopause. In their thirties sufferers need lens transplants to cure cataracts, and their arteries stiffen and become covered in fat deposits. In their forties they die, usually from heart attacks.

Werner's syndrome is one of a group of inherited rapid-ageing disorders called 'progerias'. The disorder is caused by mutations that disable a protein that maintains the integrity of DNA during replication. Cells that lack the protein have very high mutation rates. This barrage of mutations causes the cells to die instead of proliferating, or else to produce abnormal proteins. Tissues, such as skin, which rely on large numbers of dividing cells in order to maintain their integrity, fall apart. Perhaps something similar happens to us all, only at a much slower rate.

As we age, vitality slips away from our cells. This can be seen in the laboratory. It has long been possible to grow human cells in petri-dishes by means of elaborate and delicate protocols. No matter how salubrious their environment, however, freshly harvested cells will divide only a certain number of times and then divide no more. Their decline is gradual, and is caused by some intrinsic limit. Many have suggested that this cellular senescence is not merely a consequence of the ageing body but its direct cause.

Supporting this idea, cells taken from human foetuses can divide for about twice as many generations as can those from ninety-year-olds before sinking into decline. Perhaps, then, elderly people have many cells that are closer to the end of their replicative lifespans and which are, therefore, unable to contribute to repairing the wear and tear of everyday life as well as they might. When, therefore, in 1998, the molecular cause of the limit to cell division was discovered, and then broken, the thrill was tangible. If cellular senescence could be cured, perhaps so could ageing.

Each time a cell divides, its chromosomes must be replicated as well. But the enzymes that replicate chromosomal DNA are unable to replicate the ends of the chromosomes. These ends are, therefore, protected by sequences, thousands of base-pairs long, called telomeres that are gradually whittled away over the course of many cell divisions at a rate of about a hundred base-pairs per cell division. When the telomeres are gone, the cell can no longer divide and it dies. It is the rate of whittling that sets the fundamental clock of ageing. Or so the argument goes.

What is needed, then, is a way to prevent the attrition of telomeres. Not all cells lose their telomeres. The germ cells that give rise to eggs and sperm possess a complex enzyme called telomerase that maintains their telomeres and so confers upon them the immortality that they must necessarily have. The loss of telomeres that occurs in the rest of the body's cells is precisely due to the fact that they do not contain this enzyme. If telomerase is engineered into cells that normally lack the enzyme, their telomeres are preserved division after division. The cells also became immortal.

If the route to cellular immortality is so easy, why have we not taken it? The reason is quite simple: immortality is a property of cancers. Nearly all tumor cells have, somewhere in their history, undergone mutations that cause them to have telomerase where other cells do not. The absence of telomerase in our cells is probably one of the first defences we have against the multiplication of rogue cells. Besides, there is still little to show that short telomeres do, in fact, cause ageing. Only one experiment has addressed the problem directly: an experiment in which telomerase-defective mice were engineered and then bred for six generations.

Mice, it seems, can get by without telomerase for at least a while. The first generation of telomerase-defective mice that was ever produced showed no signs of premature ageing. In a way this is not surprising. These mice had telomeres as long as those of any other mice, for mice, like us, inherit their telomeres from their parents, and their parents were normal. For want of telomerase in their germ cells, however, each successive generation of these mutant mice started life with ever shorter telomeres. The effects became apparent by the fourth generation when the male mice proved to have few viable sperm. By the sixth generation they had none at all. Females were not sterile, but they produced fewer eggs than normal, and those they did produce often gave rise to defective embryos. By the sixth generation, too, male and female mice alike began to age prematurely. Like humans, mice go bald and grey with age, and the sixth-generation mice did so while still young.

These results provide at best mixed support for the idea that a want of telomeres causes ageing. Sufficiently short telomeres can clearly cause premature ageing; but since this happens only after six generations of attrition, they cannot be the cause of normal ageing in mice. While it is tempting to dismiss the whittling away of telomeres as an explanation of ageing in humans, it is probably too soon to do so. Laboratory mice have extraordinarily long telomeres – far longer than ours. If our telomeres are rather short at the start of our lives and must, by virtue of our greater size and longevity, undergo far more attrition than a mouse's, it remains quite possible that they matter to us.

One way to prove the point would be to clone a human. Clones should start life with abnormally short telomeres, for they are produced without the aid of germ cells and so their telomeres are never renewed. Successive generations of clones should have shorter and shorter telomeres and age with increasing rapidity – all the more so if the clone-donors are elderly. What with the global ban on human cloning this experiment is not likely to be carried out soon – unless by UFO cultists or renegade Italian obstetricians. But, of course, it has been done in animals. Sheep 6LL3, a.k.a 'Dolly', got her chromosomes from the udder-cells of a six-year-old Finn Dorset. She therefore began life with substantially worn-down telomeres. Many thought that she would age fast. Some arthritis aside, however, she was quite healthy; there was nothing exotic about the viral disease that prompted her euthanasia at the age of six. Clones of other animals such as cattle and mice often suffer from a variety of health problems such as obesity, but none have been reported to be progeric. Still, these are early days.

Telomerase-mutant humans would be informative too. There is another progeria, rarer than Werner's but even more severe, in which catastrophic ageing begins in childhood. The victims of this disorder usually die by the age of twelve or so, again from heart attacks, by which time they are to all appearances very small octogenarians. Their symptoms suggest defective telomeres. Even if this grim disease can be explained by too-rapid cellular senescence, we will have penetrated only a small way into ageing's mysteries. For while the progerias hasten some aspects of physical decline, they leave the minds of their victims untouched.

MAKING A CENTURY

In the last ten years there has been a revolution in the study of ageing. Much of it has come from the study of the nematode worm *Caenorhabditis elegans*. This worm is only about 1 millimetre long, and it is possible to grow thousands of them in petri-dishes. They are perfectly transparent. Under a powerful microscope it is possible to see every single one of the 959 cells in their living bodies. For whatever reason, it has been

especially easy to identify worm mutants that are extraordinarily long-lived. Some of these mutant worms live twice as
long as normal worms do: forty-two days – in human terms,
about 150 years.

So far, at least a hundred genes have been identified in worms
that, when mutated, cause them to live longer. Many of these
mutations disable the worm's insulin-like growth-factor-
signalling pathway. As a consequence of doing so, the whole
physiology of the worm changes. Mutant worms that are defective for IGF signalling reproduce less, store large amounts of fat
and sugars, and activate a whole battery of genes that encode for
stress-resistance proteins, among them superoxide dismutase.
The result is worms that radiate health even as their normal
contemporaries wither in their petri-dishes.

We have come across insulin-like growth factor before. It is
the lack of this hormone that makes pygmies small and its excess
that makes Great Danes large. It is also one of the hormones
that, when inactivated in mice, cause them to be dwarf and long-
lived. In worms, IGF does not seem to control body size (something of a surprise since it does so in so many other creatures,
including fruit flies). Even so, taking these findings from worms
together with what is known about IGF in mice, flies and many
other creatures, it is possible to sketch an account of a mechanism, perhaps universal to all animal life, that allows animals to
live longer when they need to.

Worms are not frightfully bright. The nervous system of any
one worm, including what passes for its brain, contains only 302
neurons; a human brain has around a billion-fold more. Even so,

a worm has nous enough to know how much food it has. When a worm perceives that it is about to starve, neuronal signals from sense organs in its head signal the rest of the body and IGF signalling is shut off. A change in environment mimics what many mutants do, and the result is the same: the worm lives longer.

This should sound familiar. It is, in effect, what happens in caloric restriction in mice and rats. And it suggests an interpretation for how and why *la vita sobria* has its beneficial effects. Far from being an odd laboratory phenomenon of interest only to gerontologists and diet gurus pursuing dreams of immortality, the caloric restriction response is probably a device that has evolved to allow animals to cope with the vicissitudes of life. Perceiving that it is in for hard times, a young animal alters its mode of life. Instead of investing resources in growing large and reproducing soon, it switches to survival mode. It remains small and ceases to reproduce, in effect gambling that sooner or later better times will come. If this view of caloric restriction is correct, then its enthusiasts are attempting nothing less than the revival of devices evolved to cope with the deprivation that was surely our lot for millennia of prehistory (and surely a lot of history too). Though they do not know it, when they calculate their foods to the last calorie, surround themselves with bottled vitamins, and monitor, as they must, their bone density by the month, they are playing the part of civilisation's most dedicated discontents.

Can longevity genes be found in humans? Many scientists think so. In France, Britain, Holland, Japan, Finland and the United States gerontologists are busily compiling lists of centenarians and analysing their DNA in order to find out why they live so long. They do so not in the expectation that there is any one mutation or polymorphism that all these centenarians have in common – and they fully accept that some centenarians will have made their century by a combination of good luck and virtuous living. Rather, the approach is to scan many genes which, for one reason or another, are believed to contribute to the diseases of old age and to search for those variants that are more common in geriatric survivors relative to the rest of the population.

One of the first longevity genes to be identified in this way was apolipoprotein E (APOE). The protein encoded by this gene comes in several polymorphic variants called ε2, ε3 and ε4. About 11 per cent of Frenchmen and women under the age of seventy carry at least one copy of the ε4 variant, but in French centenarians this number drops to 5 per cent, the difference being made up by the ε2 variant, which becomes more common. This implies that should you wish to see your hundredth birthday, you should hope to have at least one copy of ε2 but none of ε4.

This is because the APOE gene, which encodes a protein involved in cholesterol transport, has been implicated in Alzheimer's disease. About one in ten people aged sixty-five or over will contract Alzheimer's, but the odds are skewed drasti-cally if you are an ε4 carrier. One copy of ε4 relative to none increases your risk of Alzheimer's three-fold; two copies increases your risk eight-fold. Were this not enough, ε4 also predisposes to cardiovascular disease. With this sort of molecu-lar double jeopardy it is easy to see why ε4 carriers rarely survive to a great age.

All this seems to matter less if you are black. Surveys of APOE genes have shown that ε4 is very common in sub-Saharan Africa. Nearly half of African pygmies carry at least one copy. Does this really mean that Alzheimer's disease is rampant among the Efe? The short answer is that we don't know. No studies on the epidemiology of Alzheimer's seem to have been carried out on pygmies, and they would be hard to do since a high rate of death due to infection and accidents means that few pygmies survive to an age when Alzheimer's might be seen. This, in itself, may explain why ε4 is so common among them, but a more likely explanation is that it is less dangerous to Africans than it is to Europeans. Several studies have sought, and failed, to find an increased risk of Alzheimer's in Nigerians and African Americans who carry the ε4 variant. Why this should be so is something of a mystery.

In Europeans, at least, the genetics of Alzheimer's provide a beautiful illustration of the evolutionary theory of ageing that is, if anything, even more persuasive than that of Huntington's. Even among the clearly susceptible (white) French, ε4 is common for such a lethal variant, and its presence can only be explained by the fact that it has little net effect on the reproductive success of its carriers. The contrast with other genes that cause Alzheimer's is instructive. Mutations in at least three other genes cause Alzheimer's but do so at around age thirty. They kill their carriers in their prime and, exposed to the full force of natural selection, are accordingly – thankfully – rare.

These kinds of findings are only the beginning. Within a few years, dozens, if not hundreds, of polymorphisms will be found that add years to our lives or else take them away. Most of these polymorphisms will either hasten or else delay the features of ageing with which we are familiar: senile dementia, arteriosclerosis, kidney failure, prostate failure, menopause, cancer and the like. No single person's genome will possess all the variants that might be desirable for long life. This much is already apparent from the sheer diversity of ways in which we die. But it will be possible to describe in actuarial terms the relative risk of possessing a given genome. Here is a taste of what is to come. All else being equal, a forty-year-old whose genome has the following variants:

SRY(−/−); APOE(ε_2/ε_2); ACE(D/D); MTHFR(Ala222/Ala222)

will have a lower risk of cardiovascular disease, and hence a lower yearly risk of death, than someone with the following:

SRY(+/−); APOE(ε_4/ε_4); ACE(I/I); MTHFR(Val222/Val222).

The difference between these two lists is quite unmysterious. There are four genes – SRY, APOE, ACE and MTHFR – each of which has two variants known to be associated with a difference in the mortality rates of middle-aged or elderly people. These two lists are, then, a predictive theory of longevity, but one that is no more profound than the assertion than someone who neither smokes, drinks, drives or has sex will generally live longer than someone who does all those things. Only here the risk factors lie in the genome.

Possession of the second genome does not inevitably spell an early death. While it is not possible to diet your way out of Alzheimer's, much can be done to prevent a heart attack. That these genes confer different risks of death at any given age seems certain, but it is not yet possible to translate those differences into years. To do so requires large population studies of a sort that have not yet been done, but that surely will. There is one exception to this. In the USA, SRY(–/–) individuals live, on average, five years longer than those who are SRY(+/–). This, of course, is rather hard on those of us who are SRY(+/–), but there's not a lot that can be done about it except to give a Gallic shrug and mutter *Vive la différence*.

EVER UPWARDS

In 1994 a remarkable thing happened. Not a single eight-year-old Swedish girl died. Not one succumbed to the 'flu; not one was hit by a bus. At the beginning of the year there were 112,521 of them. At the end of the year they were all still there.

It was, of course, a statistical fluke. In that same year some eight-year-old Swedish boys died, so did some seven- and nine-year-old girls, and a few eight-year-olds of both sexes died the following year. But the survival, in that year, of those Swedish girls may be taken as symbolic of the greatest accomplishment of industrial civilisation: the protection of children from death.

Childhood mortality rates in the most advanced economies have become vanishingly small, particularly when death due to accident or violence is excluded. It is this accomplishment, at least 250 years in the making, which has driven the long climb in human life expectancy. Before 1750, a newborn child could expect to live to twenty years of age; today in the wealthiest countries a newborn can expect to live to about seventy-five. Most of this increase can be credited to the elimination of infectious diseases that preferentially strike the young. The curious thing, however, is that even though the protection of the young is largely a completed project – in the wealthiest countries – life expectancy continues to rise.

The 1960s were, it is said, revolutionary. But far from the *Sturm und Drang* of the cultural and sexual revolutions, something far more important was happening. Mortality rates of the old began to decline. An American woman who turned eighty in 1970 had a 30 per cent chance of surviving another decade; had she turned eighty in 1997 her chance of doing so would have increased to 40 per cent. The same phenomenon can be seen in the progress of maximum longevity in Sweden. Between 1860 and 1960, the age at death of Sweden's oldest person increased steadily, decade by decade, at rate of about 0.4 years. Between 1969 and 1999, the rate of increase climbed to about 1.1 years per decade. We have been living longer for some time, but since the 1960s we have been living longer ever faster.

These numbers tell us that not only can ageing be cured, but that cures have been coming thick and fast. If ageing is the age-dependent increase in the mortality rate, then anything that ameliorates the mortality rate is, by definition, its cure. The decline of mortality rates among the old is mainly due to a several-decades-long decline in cardiovascular disease and cancer. Cardiovascular disease has been the leading cause of death in the United States since the 1920s, but between 1950 and 1996 its contribution to the death rate declined by half. In Japan, cancer rates began to decline in the 1960s; in the rest of the industrialised world the decline began about twenty years later. Nothing spectacular, then, just the incremental advance of public health.

But incremental advance is all we can reasonably expect. Evolutionary theory and the increasing flow of information about the genetics of ageing, be it premature or postponed, tell us that ageing is many diseases that will have to be cured one by one. At the same time, there is no obvious impediment to that advance; nothing to make us think that human beings have a fixed lifespan. In 1994, 1674 eighty-year-old Swedish women died. It is impossible to predict what medical breakthroughs will be required to ensure that none will die in the future. But when that day comes, it will mark the completion of industrial civilisation's second great project: the protection of the old from death.

X

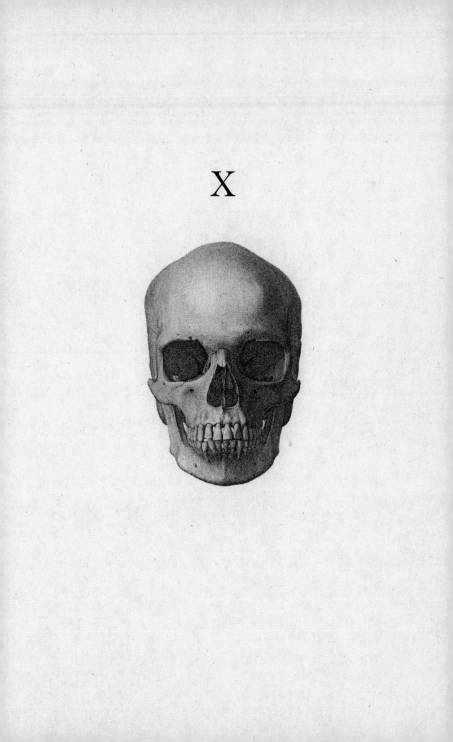

ANTHROPOMETAMORPHOSIS

[An Epilogue]

THE AUTHORS OF BOOKS ABOUT GENETICS — at least books written for the general reader — disagree about many things. What they agree on is the need either to predict the future course of humanity or to moralise about it, or, better yet, to do both. The predictions invariably concern the role that the 'new genetic technologies' — mass genetic screening, embryo selection, cloning, germ-line modification and the like — will have in the lives of individuals and societies. Some writers are sanguine, and assure us that these technologies are as nothing compared to the great demographic forces such as birth rates and migrations that shape the gene pool of our species; others contemplate, with surprising equanimity, the transformation of the human species into something resembling a highly intelligent

SKULL OF AN AUSTRALIAN ABORIGINE, ARNHEM LAND. FROM ARMAND DE QUATREFAGES 1882 *Crania ethnica: les cranes des races humaines.*

plant. Some – media scientists usually – would like to clone themselves; others would like to restructure the tax system so that people with inherited disorders have a disincentive to reproduce. Others again – *soi-disant* ethicists, dialectical biologists and bishops – speak portentously of 'human dignity', rail against 'genetic determinism' or else mutter darkly about the 'ethical dilemmas that face us all' – though rarely condescend to explain exactly what these might be.

I propose to resist the temptation to do any of this. I do not know the future, and my views on the morality of cloning humans, or germ-line engineering, or any other topic, are neither so deeply considered nor so unique as to warrant public exposure – a reticence born not of cowardice, but rather of courtesy, or so I like to think. Instead I will end this book, which has been all about the many things that we know about the construction of the human body, with some thoughts on what we do not know and what we – or at least I – would like to.

I would like to know about variety. Most of this book has been about the rare mutations that damage the body. If I have mentioned variety, I have done so only in passing. By variety I mean the normal variation in human appearance and attributes that we see in healthy people around us. I mean the variety that can be found within the smallest Scottish hamlet, with its brown-, green- and blue-eyed inhabitants. But I also mean the differences in form between populations of people who live near to each other, but are somehow distinct: short pygmies versus taller Bantu farmers, for example. And I also mean the differences in skin colour, hair curliness and eye shape that distinguish – more or less – people who originate from different continents. One of the things, then, that I want to know about is race.

Race has long been under siege. Among scientists, geneticists have led the assault. Their attack has been predicated on two empirical results that have emerged from the study of patterns of genetic variation across the globe. The first was the discovery that most of the variety so abundantly visible in our genomes does not divide humanity along lines that correspond to the races of traditional and folk anthropology. All genes come in different variants, even if most of those variants are 'silent' and do not affect the structure of the proteins they encode. Inevitably, some variants are more common in some parts of the world than others. But the ubiquity and rarity of most variant genes across the globe do not correspond to traditional racial boundaries. Racial boundaries are usually held to be sharp; gene variant frequency changes are generally smooth. Changes in variant frequencies are also inconsistent between one gene and another. If there are lines to be drawn through humanity, most genes simply don't show where they should go.

The second discovery that caused, and causes, geneticists to doubt the existence of races is the ubiquity of genetic variation within even the smallest populations. About 85 per cent of the global stock of genetic variation can be found within any country or population – Cambodians or Nigerians, say. About another 8 per cent distinguish nations from each other – the Dutch from the Spanish – which leaves only a paltry 7 per cent or so to account for differences between continents or, in the most generous interpretation of the term, 'races'. To be sure, there are genetic differences between a Dutchman and a Dinka, but not many more than between any two natives of Delft.

These facts about human genetic variation have been known since the 1960s. In each decade since then they have been confirmed, with ever more lavish quantities of data, using ever more sophisticated means of finding and analysing human genetic variation. In the 1960s genetic variation was studied by examining the migration of variant proteins on gels; today it is studied by sequencing entire genomes. Generations of scientists have expounded these results much as I have here – and asserted that, as far as genetics is concerned, races do not exist. They are reifications, social constructs, or else they are the remnants of discredited ideologies.

Most people have remained unconvinced. They have absorbed the message that races are, somehow, not quite what they used to be. Far better, then, to avoid the word and substitute 'ethnicity' or some similar term that comfortably conflates cultural and physical variety. For some, the persistence of the idea of race is a sign of racism's tenacity. I doubt that this is true. Instead I suspect that the reason the lesson of genetics has been so widely ignored is that it seems to contradict the evidence of our eyes. If races *don't* exist, then why does a moment's glance at a stranger's face serve to identify the continent, perhaps even the country, from which he or his family came?

The answer to this question must lie in that 7 per cent – paltry though it is – of global genetic variation that distinguishes people in different parts of the world. Seven per cent is a small part of global genetic variation, but it is large enough to imply the existence of hundreds, perhaps thousands, of genetic polymorphisms that are common, even ubiquitous, on one continent but rare, or even absent, on another. In recent years some geneticists have begun searching for such variants. The variants are known as AIMs or 'Ancestry Informative Markers' – so called because they can indeed tell you roughly where your ancestors came from, and even sort them out if – the case for so many of us – they came from several different places. The search for AIMs, which initially focused on Africans and Europeans but is already being extended, is prompted by the hope of identifying the genetic basis of several diseases, among them type 2 diabetes, the risk of which differs among Africans and Europeans.

Many AIMs have already been found. For some of them, the reasons for their presence on one continent but absence on another is readily apparent. One variant of the FY gene is ubiquitous in Africans but extremely rare among everyone else. The African variant is odd insofar that it prevents the protein that FY encodes from being made. Everyone else makes the protein, though its exact form can vary too. The FY protein is a growth-factor receptor found on blood cells, one that the malaria parasite seems to use as well, and its absence in Africans is almost certainly the result of long-standing natural selection for resistance to the disease. The absence of FY in Africans and its presence everywhere else has been known for decades. Many other differences are now being found – although they are usually not as dramatic as FY's. No one knows what most AIMS do or why they are there.

Somewhere among all those AIMs, however, will be the genes that give a Han Chinese child the curve in her eyelid and a Solomon Islander his black-, verging on purple-, coloured skin. Among them, too, will be the genes that affect the shape of our skulls. Skull measuring has a long history in anthropology. One of the first really assiduous skull measurers was a Dutchman, Petrus Camper, who in the 1700s invented the 'facial angle' – essentially an index of facial flatness. In his most famous diagram, Camper shows a series of heads and skulls – monkey, orangutan, African, European, Greek statue – with ever-declining facial angles. Camper himself was no racist. In his writings he emphasised repeatedly the close relationship between all humans no matter what their origins. 'Proffer with me,' he urged in 1764, 'a fraternal hand to Negroes and recognise them for veritable descendants of the first man, to whom we all look as our common father.' To which he added that the first man may have been white, brown or black, and that Europeans are really just 'white Moors' – and did so at a time when Linnaeus was carving up our species.

Sadly, Camper's iconography spoke louder than his words, and his diagram with its implicit demonstration of a hierarchy from ape to Apollo (with Africans rather closer to apes than to gods) became a staple of nineteenth-century anthropology. There is no need to recap and critique the craniometric studies carried out in the nineteenth and early twentieth centuries that sought to demonstrate that one subset of humanity was more or less intelligent than another – others have done so with a thoroughness that their scientific influence scarcely merits. But it is

340

worth noting that modern physical anthropologists remain keen on describing skull shape, though nowadays they tend to do so with 3-d laser scanners and multivariate statistics. They find, perhaps unsurprisingly, that for all the variety within populations, people from different parts of the world have different-shaped heads.

Much as Camper claimed, the jaws of sub-Saharan Africans *do* protrude, on average, further from their foreheads than do the jaws of Europeans – an attribute known as 'prognathism'. Melanesians and Australian Aborigines are also more prognathic than Europeans. *Contra* Camper, however, this does not make African (or Aborigine) skulls more like ape skulls than European ones. The facial angle is a rather crude way of capturing an exceedingly complex aspect of skull shape. It does not discriminate between different ways of being prognathic. A chimpanzee has a high facial angle because its whole face and forehead slope; Africans and Aborigines have slightly higher facial angles than Europeans because of a jut in the jaw alone. Besides, Europeans do not even have the flattest faces. That honour – if honour it is – must go to the Inuit of northern Canada.

Human skulls are wonderfully diverse. The Inuit are also notable for the largeness of their eye orbits and the massiveness of their cheekbones. Compared to everyone else, the Khoisans of southern Africa have bulging foreheads (frontal bossing); Australian Aborigines have massive brows (supra-orbital ridges); some sub-Saharan Africans have widely set eyes (large inter-orbital distances); Andaman Islanders (negritos) have small, round skulls – the list of differences could be extended

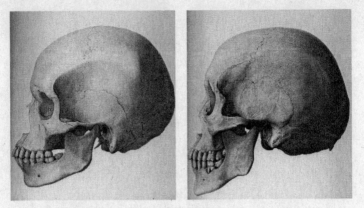

VARIATION IN HUMAN SKULLS: LEFT TO RIGHT: AUSTRALIAN
ABORIGINE, CHINESE, EUROPEAN AND KHOISAN.
FROM ARMAND DE QUATREFAGES 1882 *Crania ethnica:
les cranes des races humaines.*

indefinitely. Few of these differences are absolute. Just as most of
the variance in gene frequency is found within, rather than
among, populations (nations, continents), so too is most of the
variance in skull shape. And the differences among populations
are all subtle. Australian Aborigines and Inuit differ in prognathism by only 6 per cent. Small differences, then, but differences that, given the attention we devote to each other's faces,
strike us immediately.

My claim that we will soon be able to identify the genes responsible for all this diversity in skull shape suggests an important
question: namely, do such genes exist? In 1912 the American
anthropologist Franz Boas set out to demonstrate that they do
not. A humane and tolerant man, he was an implacable opponent
of those who sought to make invidious distinctions between
humanity based on the shapes of their skulls. The following

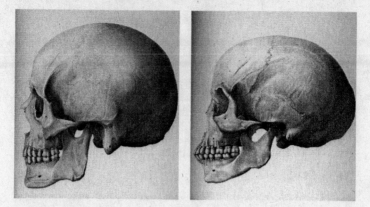

passage, taken from a serious anthropological article written in 1905 by a German dentist called Röse, gives a flavour of what he was up against: 'The long heads of German descent represent the bearers of higher spiritual life, the occupants of dominant positions, to which they are destined by nature, the innate defenders of the fatherland and the social order. Their whole character pre-determines them to aristocracy.' And so on, to the detriment of the more democratically minded and un-German 'round heads'.

The 'long' and 'round' heads refer to the value of the 'cephalic index', the ratio of skull breadth to width (expressed as a percentage, long heads or dolichocephalics have a cephalic index below seventy-five, while round heads or brachycephalics have a cephalic index above eighty; mesocephalics are somewhere in between). Noting that the immigrants who arrived at Ellis Island – Bohemians, Slovakians, Hungarians, Italians, Scots and Eastern European Jews – varied somewhat in their cephalic indexes, Boas asked whether these differences were due to genetic (to use his terminology, 'racial') or environmental causes. He reasoned, soundly, that if the skulls of the American-born children of all these various groups were more similar to each other than those of European-born children, then environment rather than ancestry must be the cause of the differences. Boas measured some thirteen thousand heads – a vast undertaking that left him, in the absence of computers, overwhelmed by numbers. Nevertheless, he managed to produce a graph that seemed to show that the cephalic indices of the US-born children of Sicilians and Eastern European Jews (both rather dolichocephalic to begin with) were, indeed, converging. It was a case of new heads for the New World.

Boas's study dealt a near-fatal blow to craniometry. Over the last ninety years it has been cited innumerable times – not least by the late Stephen Jay Gould – as proof that skull shape is 'plastic', that is, caused by non-genetic differences. Boas, however, was wrong. His data have recently been comprehensively re-analysed using modern statistical techniques. The skulls of American-born children do indeed differ from those of their parents, but they do so inconsistently. Indeed, had Boas chosen to compare the children of Scots and Hungarians, rather than those of Sicilians and Eastern European Jews, he could have shown that America causes skulls to diverge rather than converge in shape. But he was wrong in a deeper sense than this. Re-analysis of his data also shows that the changes in skull shape caused by American birth, whatever their direction, are trivial compared to the differences that remain and that are due to ancestry and family – or, to put it more succinctly, to genes. Indeed, looking beyond European immigrants, this is hardly surprising. Forensic anthropologists in the United States and Britain are quite adept at telling whether a given skull, perhaps evidence of some foul deed, once belonged to someone of African or European ancestry. That they can do so after decades, even centuries, of co-existence, not to mention generous amounts of admixture, suggests that our differences are not, as is often said, merely skin deep, but extend to our skulls – if not to what they contain.

So, genetic differences exist among all sorts of people. Should we try to find out what they are? Many scientists think not. Some find it enough to dismiss such physical diversity as exists among human populations as 'uninteresting' – not worthy of study. Others concede that it may be interesting, but that it should not be studied, since even to contemplate doing so is to engender social injustice. They fear a revival of not merely racial, but racist, science.

For my part, I should love to know the genes responsible for human diversity; the genes for the differences – be those differences between men and women who live in the same village or those who have never trodden on each other's continents. In part this is simply for the pleasure of knowing. This the pleasure that comes from looking at Gabriel Dante Rossetti's painting *La Ghirlandata* and knowing that his model, Alexa Wilding, had two loss-of-function MC1R mutations that gave her such glorious red hair. This pleasure of knowing is partly that which all science gives, but to which is added the pleasure that comes from understanding the reason for something that has been hitherto at once familiar but completely mysterious.

The view that human diversity is dull seems to me excessively Olympian. After all, if population geneticists have ignored human variety, they have for decades lavished their (seemingly inexhaustible) energies studying variety in the colours of garden-snail shells and the number of bristles that decorate the backs of fruit flies – problems that are intellectually much like those presented by human variation.

The claim that human genetics is morally dangerous is a more serious one. One can certainly, given the history of racial science, see where such a claim originates. Nevertheless it is misplaced. Reasonable people *know* that the differences among humans are so slight that they cannot be used to undermine any conceivable commitment to social justice. 'Human equality,' to borrow a slogan of Stephen Jay Gould's, 'is a contingent fact of human history.' What is true, however, is that as long as the cause of human variety remains unknown – as long as the 7 per cent of genetic variance that distinguishes people from different parts of the world remains obscure – there will always be those who will use that obscurity to promote theories with socially unjust consequences. Injustice can sometimes be the consequence of new knowledge, but more often – far more often – it slips in through the cracks of our ignorance.

Perhaps the most compelling reason that we should once again turn our attention fully to the study of human physical diversity is that it is disappearing. In South-East Asia the negritos, those enigmatic pygmy-like people, are in decline. They are hunter-gatherers. Overrun by Austronesian-speaking farmers in the Neolithic, they mostly persist on remote islands. Now, modernity threatens. On Lesser Andaman, the remaining Onge live in reservations. On Greater Andaman, a few hundred

Jarawas survive by virtue of having fended off the curious with bows and arrows (in the last fifty years they have killed or injured more than a hundred people), but they too have now emerged from the forest, attracted by baubles offered by Indian officials. It is feared that they will soon succumb to tuberculosis, measles and culture shock as their predecessors have.

They are only the latest casualties of Austronesian and European (not to mention Chinese, Bantu and Harappan) expansion. In 1520 Ferdinand Magellan, arriving at the straits that today bear his name, reported the existence of a race of giants that lived in the interior of Tierra del Fuego. He called them the *Pataghoni*, after a giant in a Spanish tale of chivalry. Subsequent travellers embroidered the account; by 1767 these giants, a wild and brutal people, had grown to about three metres (ten feet) tall. Today, the giants of Tierra del Fuego are as forgotten and fantastical as Pliny's Arimaspeans. And yet the *Pataghoni* existed. They called themselves the Selk'nam or Ona, and they had an average adult-male height of 178 centimetres (five feet ten inches) – giant, then, but only to sixteenth-century Spanish sailors. But if their stature was not that remarkable, their skulls certainly are. They have a strength and thickness, a robustness, which other human skulls don't, and this is true of their skeletons as a whole. Some photographs of the Selk'nam exist. They depict a handsome and physically powerful people who wore cloaks made from the pelts of the guanacos that they hunted on foot using bows as tall as themselves. Argentine sheep ranchers killed the Selk'nam off in a genocidal slaughter, and the last one died some time around 1920.

GROUP OF SELK'NAM, TIERRA DEL FUEGO C. 1914

There is one more thing I should like to know about. And it is a phenomenon more general and nearly as contentious as race. It is beauty. Beauty is that which we see (or hear or touch or smell) that gives us pleasure, and as such its forms are, or at least seem to be, infinitely various. Here I am concerned with physical beauty alone.

'Beauty,' says the philosopher Elaine Scarry, 'prompts the begetting of children: when the eye sees something beautiful, the whole body wants to reproduce it.' Plato, she points out, had the same idea. In *The symposium*, Socrates tells how he was instruct-ed in the arts of love by Diotima, a woman of Mantinea, and how they spoke of the nature of love and beauty. 'I will put it more plainly,' says Diotima. 'The object of love, Socrates, is not what you think, beauty.' 'What is it then?' 'Its object is to pro-create and bring forth beauty.' 'Really?' 'It is so, I assure you.'

348

Darwin could not have put it better himself. Much of his *The descent of man and selection in relation to sex* is devoted to investigating the presence, perception and purpose of beauty. 'The most refined beauty,' he wrote, 'may serve as a charm for the female, and for no other purpose.' He was thinking of the tail feathers of the male Argus pheasant with its geometrical arrays of ocelli. But the psychologies of pheasants and Fijians are really much the same. For Darwin, the love of beauty is a very general evolutionary force, second only to natural selection itself in power. Creatures choosing beauty for generation upon generation have given the natural world much of its exuberance. Sexual selection has given the Madagascar chameleon its horns; it has given the swordtail fish its sword and Birds of Paradise and Argus pheasants their tails; it has given the human species its variety.

One of the fascinating things about Darwin's account of beauty is that without reference to philosophers or artists he stakes out a position on the great issues of aesthetics. He wants to know whether or not beauty is universal or particular, whether it is common or rare, and whether or not it has meaning. To all of these questions Darwin has an unequivocal answer. Physical beauty, he asserts, is not universal, but rather particular. Different people in different parts of the world each have their own standard of beauty. And it is rare. To be beautiful is to be a little different from everyone else around us. It is also meaningless. Our brains, for whatever reason, perceive some things as beautiful, and do so regardless of the other qualities that those things may have. Beauty does not signify anything. It exists for its own sake.

Darwin's views on beauty are characteristically, effortlessly, original. *The descent of man* contains nothing, for example, about the classical ideal of beauty – the ideal that, from Archaic kouros to Antinous' scowl, was replicated across the Mediterranean for centuries as if there were a formula for it; which there was, one that by the Renaissance had become a theory of human beauty in which proportions were divine, a theory that in the eighteenth century turned into the standard by which all humanity was judged. It was this ideal that caused Winckelmann to assert that the ancient Greeks were the most beautiful of all people (though he thought modern Neapolitans comely as well); that caused Camper to place the head of a Greek statue at one end of his continuum of facial angles; Buffon to identify a 'beauty zone' between 20 and 35 degrees north that stretched from the Ganges to Morocco and took in the Persians, Turks, Circassians, Greeks and Europeans; and Bougainville, when he arrived in Tahiti in 1768, to eulogise its inhabitants in terms of a classical idyll painted by Watteau. Darwin avoids all this. He does not tell us what he thinks is beautiful; instead he attempts to find out what other people think. He collects travellers' reports. American Indians, he is told, believe that female beauty consists of a broad, flat face, small eyes, high cheekbones, a low forehead, a broad chin, a hook nose and breasts hanging down to the belt. Manchu Chinese prefer women with enormous ears. In Cochin China, beauties have round heads; in Siam, they have divergent nostrils; Hottentots like their women so immensely steatopygous that, having sat down, they cannot stand up again.

Darwin does wonder about the quality of his data – and

rightly so. But in general he is quite convinced that different people perceive beauty in different ways. His vision is an appealing one. *Per molto variare la natur è bella* – nature's beauty is its variety; it could be Darwin's slogan and it could be Benetton's (though it was Elizabeth I's). Indeed, when we consider the whirlwind of fashion it is impossible to doubt that the love of beauty is frequently the particular love of rare and meaningless things. Among scientists who study beauty, however – and the study of beauty is itself increasingly fashionable – Darwin's views are seen as rather quaint. These days, most research on the subject begins with the notions that the standard of beauty is universal, that the presence of beauty is rather common, and that far from being meaningless, it has a great deal to say.

The universality of beauty's standard is as self-evident as its particularity. The apparent contradiction is resolved if we simply recognise that there are some things about which tastes differ, and some about which they do not. Tastes in hairiness (cranial, facial, bodily), pigmentation (eye, hair and skin colour) and perhaps even body shape (hip–waist ratio) all seem to differ quite a lot from person to person, place to place, time to time. But the taste for relative youth – at least when men judge women – does not. Nor, it seems, does the taste for certain kinds of faces. Average faces seem to be universally more attractive than most, but not all, variant faces. Symmetry is preferred over asymmetry. These are some of the results of a large literature devoted to finding out who finds what beautiful when. Much of it demonstrates the obvious. After all, were a Papua New Guinea tribesman brought to London's National

Gallery and offered the choice of Botticelli's Venus (she of *Mars and Venus*) and Massys's *Grotesque Old Woman* as a mate, he might well be unimpressed by either, but we can be sure which one he'd choose.

Beauty's meaning is more controversial. Here I wish to pursue just one idea: that it has something to do with physiological condition; that it is, indeed, a certificate of health. In its simplest form the truth of this idea is also quite self-evident. Clear skin, bright eyes and white teeth are manifestly signs of beauty and health. It is no accident that Brazilian men, glimpsing a beautiful carioca, sigh '*Que saúde*' – what health. Whether particular facial proportions and symmetry signify health is, however, less obvious. Studies using computer-generated faces show that we perceive beautiful faces as being healthy ones. But searches for a correlation between the beauty and health of real people have found only weak and inconsistent effects.

Perhaps this is because beauty is no longer what it was. For all of human history, poor health has mostly been about nutrition and pathogens – a lack of the first and an excess of the second. Beauty was an indicator of the salubriousness of the environment or else the ability to resist its vicissitudes. To the degree that this is true, then the variance in beauty must be declining in the most developed nations at least, even as its mean increases. Goitres and cretinism may still afflict large parts of the world, but they no longer afflict the Swiss. The scars of smallpox have disappeared everywhere. Even in England most people now keep their teeth until they die. One wonders whether the diseases – filariasis, malaria, sleeping sickness, not to mention

ANTHROPOMETAMORPHOSIS

nutritional deprivation in its many forms – that afflict so many of the world's children can be read in the symmetry and proportions of their faces if, as adults, they should have survived them. There is no doubt that prosperity exacts a cost to beauty in the form of obesity, dental cavities and stress. But if the balance of its effects is favourable, and it must be, then any classroom of American or European undergraduates contains an abundance of beauty that has never existed in human history before.

That may seem implausible, but only because we have little grasp of beauty's advance. Beauty is like wealth. It increases over time, yet its distribution remains unequal. However much of it we have, it always seems that someone else has more. In part this is because beauty, as the consequence of health, is also the consequence of wealth. But suppose there existed a society so wealthy and egalitarian that, as far as pathogens and nutrition are concerned, all were equally healthy. A society of the sort approached by the Netherlands (but from which Great Britain and the United States remain woefully distant), in which the socio-economic background of a child cannot be judged from his or her physical appearance alone. Would all be equally beautiful in such a society? Would beauty's difference have disappeared? I doubt it. However beautiful the average Dutchman may believe himself to be, some of his compatriots will be more beautiful yet. I suspect that there is a residual variance in beauty that even the most controlled upbringing cannot eradicate. A residuum that lies in our genes.

The effects of poor childhood nutrition and exposure to pathogens upon the face may be uncertain, but the effects of mutations are not. When clinical geneticists attempt to classify the symptoms that their patients present, it is to the face that they first look. They are expert in recognising the subtleties that are often the only outward sign of deeper disturbances in the genetic order: shallow philtrum, low-set ears, upturned nose, narrow or wide-set eyes. Many, perhaps most, of the disorders that I have discussed in this book – from achondroplasia to pycnodysostosis – can be read in the face.

It seems that our faces are very vulnerable to mutation. Or perhaps we are just very good at reading mutation's effects in them. Either way, it seems likely that mutation's effects are written on all of our faces – not simply the faces of people with identified clinical disorders. I began this book by observing that every newly conceived embryo has, at an educated guess, an average of three hundred mutations that affect its health for the worse. It may seem impossible that we could, as a species, be so poorly. But a certain number of mutations are eliminated by selection in the womb. A woman who knows that she is pregnant has a 15 per cent chance of miscarriage; many more embryos must be lost to women who are unaware that they have conceived. More than 70 per cent of spontaneously aborted foetuses bear severe chromosomal abnormalities, and it is likely that many also bear mutations in particular genes. It is now widely supposed that miscarriage is an evolved device that enables mothers to screen for, and rid themselves of, genetically impaired progeny.

Mutation is a game of chance, one we must all play, and at which we all lose. But some of us lose more heavily than others.

Some calculations hint at the distribution of our losses. If we suppose that, of the three hundred mutations that burden the average newly conceived embryo, five are lost from the population each generation by death (miscarriage, infant and childhood mortality), then the average adult carries 295 deleterious mutations. The least burdened 1 per cent of the population will have about 250 mutations, and the most burdened 342. Somewhere in the world there is a person who has the fewest mutations of all, about 191 of them.

These calculations confirm the intuition that no one leaves the genetic casino unscathed. But they are just educated guesses. They also take no account of the relative cost of each mutation. They are the equivalent of estimating gambling losses by counting the number of chips surrendered to the house without noting their value. It seems likely that the cost of most mutations is quite small. They give us minor ailments such as bad backs and weak eyes. I suspect that they also give us misaligned teeth, graceless noses and asymmetrical ears. If this is so, then the true meaning of beauty is the relative absence of genetic error.

There is, admittedly, very little evidence for this idea, at least in humans. Evolutionary biologists have long suspected that the peacock's tail and the red deer's roar are signals of genetic quality, and have amassed much evidence in the support of this theory, most of it weak. The mutational-load explanation of beauty is however consistent with our intuitions – or prejudices – about the distribution of beauty. If deleterious mutations rob us of beauty, they should do so with particular efficacy if we marry our relatives. Most novel mutations are at least partly recessive, and inbreeding should accentuate their negative effects as they become homozygous. There is no doubt that consanguinous marriages have a cost: the children of cousins have a 2 to 4 per cent higher incidence of birth defects than those of unrelated individuals. One wonders if such children would be judged less beautiful than their outbred peers as well. Pakistan, where around 60 per cent of marriages are between first cousins, would be a good place to look. Conversely, people of mixed ancestry, such as Brazilians, should show the aesthetic benefits of concealing their recessive mutations – *Que saúde*.

What makes physical beauty so wonderful? What enables it to take us by surprise, to prevent us from treating it with indifference no matter how saturated we are by the worlds of advertising and celebrity that have appropriated it, indeed made us suspicious of its power? If the answer that I have sketched contains any truth, then each image of a beautiful face or perfectly turned limb is not really about the subject that it appears to be, but rather what it is not. It is about the imperfections that are absent: the machine errors that arise from the vicissitudes of the womb, childhood, maturity and old age, that are written all over our bodies and that are so ubiquitous that when we see someone who appears to have evaded them, however fleetingly, we pause to look with amazed delight. Beauty, Stendhal says, is only the promise of happiness. Perhaps. But it is equally the recollection of sorrow.

ACKNOWLEDGEMENTS

I have accumulated many debts while writing this book. My agent, Katinka Matson at Brockman Inc., first saw what *Mutants* might become. I thank her as well as Karen Murphy at Viking Penguin USA, Maarten Carbo at Contact, Netherlands and, most of all, Michael Fishwick at HarperCollins UK, whose faith in the book's ultimate existence was tested but never faltered. Robert Lacey, also at HarperCollins, was a wonderful editor. My Dutch translator, Robert Vernooy, was an acute critic. Several friends and colleagues commented on part of the manuscript, among them: Austin Burt, Arnold Heumakers, Barbara van Ijzeren, Marie-France Leroi, Jan-Roelof Oostra, Corinne Pernet and Jonathan Swire. Olivia Judson, Clare Isacke, Jennifer Rohn and Alberto Saez read and commented upon it all; I do not know how to repay them.

Many friends and colleagues answered specific queries, among them: Elizabeth Allen, Alan Ashworth, Peter Beighton, Chin Chiang, François Delange, Frank Dikötter, Saul Dubow, Ademar Freire-Macias, Frietson Galis, Jill Helms, Christiane

Hertel, Annemarie Heumakers, Michael Hochberg, Beatrice Howard, Grace Ioannidou, Martin Kemp, Hannelore Kischkewitz, Deborah Posel, Liesbet Rausing, Raymund Roos and John Wilmoth. Jan-Roelof Oostra in Amsterdam and Cédric Cremiere and Jean-Louis Fischer in Paris were especially generous with their expertise in teratology and its history. Véronique Dasen in Fribourg told me about the teratology of the ancient world; Marta Lahr and Robert Foley at Cambridge showed me their wonderful collection of skulls; Yehuda Koren and Eliat Negev in Jerusalem told me about the Ovitz family in the Third Reich. I have not been able to do their scholarship justice. My pupils Anne Rigby and Sarah Ahmed told me of things that just had to go into the book; Carolyn Richardson and Irin Maier researched and translated texts. I could not have illustrated the book without the help of Miriam Guttierez-Perez at the Wellcome Library for the History of Medicine, London, and Laura Lindgren and Gretchen Worden at the Mütter Museum, Philadelphia.

My greatest debts, however, are to those around me: my colleagues at Imperial and the – sadly neglected – people in my lab; my friends – Austin Burt, Jim Isacke, Olivia Judson, Giorgos Kokkoris, Vasso Koufopanou, Michaelis and Katerina Koutroumanidis, Alexandra Meliadou, Jenny Rohn, Jonathan Swire, Liesbeth Verreijdt; and my family – Marie-France, Harry, Iracema, Joseph, most especially my parents, Antoine and Johanna. But above all it was Clare Isacke who sustained me while writing this book. It is dedicated to her with love.

NOTES

The clinical and developmental genetic literature is both vast and growing. However, I have attempted to give a guide to where the major results can be found and, occasionally, further details on particular topics. Beyond these notes, the most important source for those seeking further information about particular genetic disorders is Mendelian Inheritance in Man, an on-line database authored and edited by Victor A. McKusick and his colleagues at Johns Hopkins University, Baltimore, and supported by the National Center for Biotechnology Information, USA. It contains continually updated descriptions of each disorder, the mutations that cause them, and the clinical genetic literature. MIM can be found at http://www3.ncbi.nlm.nih.gov/Omim/. To assist those who wish to penetrate this difficult literature I give, for each syndrome and gene, the MIM numbers in bold so: achondroplasia (**100800**) is caused by mutations in FGFR3 (**134934**). Neither MIM nor this book should be used for self-diagnosis.

PROLOGUE

xiii *Genetics, to quote one popular writer.* Steve Jones, whose book *The language of the genes* (1993), HarperCollins, London, remains the best popular account of human population genetics.

xiv *On 15 February 2001.* The sequence of the human genome (International Sequencing Consortium 2001).

xv *To learn from animals.* See Gilbert (2000) p.361 for Leonardo's cow placenta and Needham (1959) p.65 for Cleopatra's alleged studies of human development.

CHAPTER I: MUTANTS

3 *We had heard that a monster had been born at Ravenna.* The monster of Ravenna has been much discussed. See Landucci (1542; 1927) pp.249–50 for a contemporary account of the monster. Jean Céard dicusses its evolution in his edition of Paré's *Des monstres* (1573; 1971) pp.153–5; Niccoli (1990) pp.35–51 its political meaning; see also Fischer (1991) pp.54–6 and Daston and Park (1998) pp.177–82. I suggest that the monster's disorder is Roberts's syndrome (**268300**), but others (Walton et al. 1993; Martinez-Frias 1993) have suggested cyclopia, sirenomelia or else hydrocephalus. All these diagnoses are guesses – which one you favour is a matter of which depiction of the monster you use, and which of its many odd features you believe are real.

6 *In the sixteenth and seventeenth centuries.* My description of Renaissance teratology is indebted to Park and Daston (1981) and Daston and Park (1998), though it has perhaps a more Whiggish flavour than theirs. See Boaistuau (1560, 2000) for a reproduction of an unusually beautiful teratological manuscript of the time, Melancthon and Luther (1523; 1823) for the Monk-calf, and Paré (1573; 1982) pp.3–4 for his list of the causes of deformity. For nineteenth-century views on maternal impressions see Gould and Pyle (1897) and Bondeson (1997) pp.144–69. For the seventeenth-century teratologists see Aldrovandi (1642); the first edition of Liceti's *De monstrorum* was published in 1616, but I have worked from the second (1634) edition, a synopsis and French translation of which is given by Houssay (1937). A brief account of Liceti's life and work is given by Bates (2001).

8 *There is a moment in time.* For Sir Thomas Browne's views on deformity in *Religio medici* (1654) see his *Works* (1904) volume 1, pp.26–7. For the shift in opinion of monsters from wrath to wonders of God see Park and Daston (1981). For William Harvey's writings on generation see *Anatomical Exercises on the generation of animals; to which are added, essays on parturition; on the membranes, and fluids of the uterus; and on conception* (1650) in his *Works* (1965). The quote, however, is from the 1653 translation by Martin Llewellyn as given by Needham (1959) p.134.

10 *It was, however, a contemporary of Harvey's.* For Bacon's division of science see Bacon (1620, 2000) pp.148–9 and 223–4. For an account of

Renaissance collections of marvels see Daston and Park (1998) pp.255–301.
Harvey, John Aubrey tells us, thought little of Bacon as either a philosopher
or a writer, but Harvey espoused very Baconian sentiments when he wrote:
'Nature is nowhere accustomed more openly to display her secret mysteries
than in cases where she shows tracings of her workings apart from the
beaten path.'

13 *Most of these people have mutations*. On-line Mendelian Inheritance in Man
lists about a thousand genes that cause phenotypic variation, be it pathological or not (e.g. brown eyes).

16 *If there is no such thing as a perfect or normal genome*. The estimate of how
many times each of the genome's base-pairs have mutated in the last generation
alone is given by Kruglyak and Nickerson (2001). The estimate of 65 per cent
of genes as having polymorphisms applies to alleles defined by non-synony-
mous polymorphisms only. Conversely, my claim that most genes have an over-
whelmingly common variant comes from the observation that 35 per cent of
genes are monomorphic, and that in the known polymorphic ones, the minor
alleles usually have a frequency below 5 per cent. Again, this applies to non-
synonymous polymorphisms only (Cargill et al. 1999; Stephens et al. 2001).

18 *Each embryo has about a hundred mutations*. Eyre-Walker and Keightley
(1999) estimate the rate of production of deleterious mutations in humans.
Their estimates are consistent with those from Cargill et al. (1999) and
Stephens et al. (2001) obtained by other means. Crow (2000) reviews the fitness effect of novel mutations.

CHAPTER II: A PERFECT JOIN

25 *The Parodis arrived in Paris*. Contemporary accounts of Ritta and
Christina Parodi are given by Anon (1829 a; b; c); Saint-Ange (1830); Janin
(1829) and Danerow (1830). Later accounts by Thompson (1930; 1996) p.84
and Bondeson (2000) pp.168–73.

26 *The first cut exposed the ribcage*. The major anatomical monograph
on Ritta and Christina is Serres (1832). É. Geoffroy Saint-Hilaire (1829) considers the girls in a small paper and I. Geoffroy Saint-Hilaire (1832–37) in
volume 3 pp.161–74 of his synoptic teratology.

27 *The oldest known depiction*. The Anatolian statue, from the Catal Hüyük
site, dates from around 6500 BC; the Australian rock carving from 3–4000
BC. For the Molionides brothers, and a more general discussion on conjoined twins in ancient Greek art, science and myth, see Dasen (1997; 2002).
For Renaissance teratologies see Paré (1573; 1982) and Boaistuau (1560; 2000)
pp.134–7. For the Montaigne quote see Montaigne (1603; 1998), and for their
intellectual context Daston and Park (1998) pp.205–7. The conflict between
Duverney and his rivals is discussed by Fischer (1991) pp.71–4 and Wilson
(1993) pp.150–9. For the intellectual context of preformationism and epigenesis see Needham (1959) chapters 3 and 4, and Pinto-Correia (1997).

32 *What makes twins conjoin?* Conjoined twins occur at a frequency of 1 in 100,000 live births; monozygotic separate twins occur at a frequency of 1 in 300 live births. For Aristotle on conjoined twins see *The generation of animals* in his *Complete works* volume 1 pp.1192–1996. See Friedman (1981) pp.180–1 on baptising conjoined twins.

33 *Until recently, the origin of conjoined twins.* For a typical medical embryology textbook account of conjoined twinning see Sadler (2000) p.155. Although most conjoined twins seem to be monozygotic (they are nearly always of the same sex) there is at least one case that has been shown, by genetic tests, to be the result of a fusion between dizygotic embryos (Logroño et al. 1997). For the sex ratios of conjoined twins see Steinman (2001 a; b). Spencer (2000 a; b; 2001) gives a detailed critique of the fission model of conjoined twinning based on the geometry of the joins. See Martin (1880) pp.153–69 for the evolution of theories of the causes of conjoined twins.

35 *On the seventh day.* For a description of early human embryogenesis see Beddington and Robertson (1999) and Sadler (2000).

37 *In the spring of 1920.* For the Hilda Mangold (*née* Pröscholdt) paper see Spemann and Mangold (1924); for a translation and commentary see Willier and Oppenheimer (1964); for her biography see Hamburger (1988) and Fässler and Sander (1996).

39 *For seventy years.* For a brief history of the search for the organiser molecules see Gilbert (2002) *A selective history of induction.* http://zygote.swarthmore.edu/. Spemann quoted in Gilbert (2000).

40 *It would be tedious to recount.* 'Noggin' is slang for 'head'. For the initial identification of noggin (**602991**) see Lamb et al. (1993); for contemporary commentary see Baringa (1993); for a textbook survey of the organiser see Gilbert (2000) pp.303–38; and for a recent technical review see Beddington and Robertson (1999). The number of molecules involved in cell–cell communication includes both signalling molecules and their receptors (International Sequencing Consortium 2001). For the antagonism between BMP4 and chordin (**603475**) and noggin see Zimmerman et al. (1996) and Piccolo et al. (1996). For the noggin-defective mouse see McMahon et al. (1998); for the noggin and chordin double-defective mouse see Bachiller et al. (2000).

45 *When Eng and Chang.* The 'two organiser' theory is sometimes called the *crowding* model to distinguish it from the *fission* and *fusion* models (J.-F. Oostra, pers. comm.). Most fusion models postulate separate embryonic discs. My model is very similar to that of Hamburger (1947). It also seems similar to that of Spencer (2000 a; b; 2001) though she is ambiguous as to whether conjoined twins arise from one or two embryonic discs. Although, as stated, the vast majority of conjoined twins have a single amnion and placenta, there is apparently evidence that *some* have two amnions or even – truly strange this – two placentas. The 'two organiser' model would not apply to such twins. For chemical induction of conjoined twins (and mono-amnion monzygotic twins) see Kaufman and O'Shea (1978).

46 *One man who thought deeply*. For a history of artificial incubation see Needham (1959) pp.22–5 and 203–4. For Geoffroy's attempts to artificially create monstrous chickens see Geoffroy Saint-Hilaire (1825); Fischer (1972) and Appel (1987) pp.121–9. See also his son's account of the influence of these experiments (Geoffroy Saint-Hilaire 1847). Étienne's major teratological work is Geoffroy Saint-Hilaire (1822) *Philosophie Anatomique des monstruosités humaines*. The classificatory work was carried on by Isidore in Geoffroy Saint-Hilaire (1832–37) *Histoire générale et particulière des anomalies de l'organisation chez l'homme et les animaux*. See Morin (1996) for a modern evaluation of Geoffroyean teratology. For some of Geoffroy's intellectual background see Appel (1987) pp.121–9; for a paen by Geoffroy to Bacon see Geoffroy Saint-Hilaire (1825).

49 *Geoffroy was deeply enamoured*. Isidore Geoffroy Saint-Hilaire (1847) claimed that his father first conceived of the *soi pour soi* in 1826, several years before Ritta and Christina appeared. A full, late statement of the law's implications can be found in É. Geoffroy Saint-Hilaire (1838); see also Appel (1987).

51 *The ability of disparate organ primordia*. For a textbook review of cell adhesion molecules see Alberts et al. (1994) pp.950–1006. Spina bifida (**182940**) and anencephaly (**206500**) are both neural defects caused by unknown genetic factors and many environmental ones (Corcoran 1998). For bifid heart see Gilbert (2000) p.474.

52 *The power of cell–cell adhesion*. Parasitic conjoined twins have an occurrence of one in a million live births. See Serres (1832) for a description of these cephalothoracoileopagus conjoined twins. See Spencer (2000 a; b; 2001) for a review of conjoined twins and parasites. Park and Daston (1981) and Bondeson (2000) pp.vii–xix tell the story of Lazarus Colloredo; see Thompson (1930; 1994) p.93 for Laloo; Ta-Mei et al. (1982) for the Chinese man and Rodriguez (1870) for the multiple parasites. Spencer (2001) persuasively argues the foetus-in-foetu-teratoma theory. Naudin ten Cate (1995) describes the twenty-one-foetus teratoma.

56 *In recent years, much has been learned*. The story of the old soldier is told by Geoffroy Saint-Hilaire (1832–37) volume 3 pp. 8–11; Martin (1880) p.147, and Fischer (1991) p.74. Geoffroy and Martin claim that the old soldier was also the inspiration for Molière's *Le Médicin malgré lui* (Doctor in spite of himself), a comedy which turned on a doctor who diagnosed a patient's heart as being on the right side of her chest. Appealing though this is, it cannot be true since Molière composed his comedy in 1666–67, Les Invalides was built in 1671, and the old soldier died in 1688 (J.-L. Fischer, pers. comm.).

56 *It is a diagnosis that allows*. Kartagener's syndrome or primary ciliary dyskinesia (**244400**) is caused by recessive mutations in the DNAI1 (**604366**) and DNAH5 (**603335**) genes (Guichard et al. 2001 and Olbrich et al. 2002). See Afzelius (1976) for the causal link to cilia; see Kosaki and Casey (1998), Casey and Hackett (2000) and Brueckner (2001) for recent reviews.

50 *I said earlier that the organiser is.* For ciliary flow around the node see Nonaka et al. (1998). Of course, in a sense the discovery of asymmetrical ciliary movement doesn't absolutely solve the problem of the embryo's handedness, since it merely raises the question of why cilia should beat from right to left rather than the other way around. There's no good answer to this question in mammals at least, but it is worth noting that the embryo is a three-dimensional object, and once two axes (head to tail and back to belly) have been established, the last (left to right) obtains automatically. So a cell that wishes to tell left from right merely needs to know which way its cilia protrude with respect to the rest of the embryo's geometry. For some conceptual models of symmetry-breaking see Lander et al. (1998).

59 *There is a lovely experiment that proves this.* The original path-breaking paper defining the signals that determine left–right asymmetry in chicks is Levin et al. (1995). Since then a large literature has accumulated on left–right signalling in all major vertebrate model systems – and it seems clear that although the precise distribution of the various signalling molecules differs between mice and chickens, the general principles are the same (Tsukui et al. 1999; Meyers and Martin 1999; Casey and Hackett 2000). The differences between mice and chickens also affect the interpretation of Levin et al. (1996) who gave the original explanation for situs inversus in conjoined twins. Their explanation, which is surely correct in essence, was based on the earlier chicken data and now cannot be right in detail.

62 *For Étienne Geoffroy Saint-Hilaire.* The quote is from his *Philosophie anatomique* (1822) and runs in full: '*les Monstres ne sont plus des jeux de la Nature; leur organisation est soumise à la loi commune; les Monstres sont d'autres êtres normaux; ou plutôt il n'y a pas de Monstres et al Nature est une*'.

CHAPTER III: THE LAST JUDGEMENT

65 *In 1890 the citizens of Amsterdam.* For a history of teratology in the Netherlands and a re-evaluation of Vrolik's specimens see the magnificent series of articles by Vrolik's successors at the University of Amsterdam: Baljet and Oostra (1998) and Oostra et al. (1998a–e). Baljet and Öjesjö (1994) have suggested that Hieronymus Bosch's demonic creatures were inspired by human malformations. Most scholars point instead to medieval illustrations or else the gargoyles of Sint Jan's cathedral, s'Hertogenbosch. Although Bosch surely cannot have seen many deformed infants in the flesh, the correspondence between many of his grotesques and known deformities is certainly striking. See also Bos and Baljet (1999).

67 *Of Willem Vrolik's published writings.* Vrolik's first major work on cyclopia was a long article in Dutch (Vrolik 1834), followed a decade later by his *Tabulae* (Vrolik 1844–49). See Baljet (1990) and Baljet et al. (1991) for an account of Vrolik's work on cyclopia and his collections.

68 *Hesiod says that there were three Cyclopes.* Homer *The Odyssey* (trans. E.V. Rieu. 1946. Penguin Books, Harmondsworth, UK); Ovid *Metamorphoses* (trans. A.D. Melville. 1986. Penguin Books, Harmondsworth, UK). For classical Cyclopean iconography see Touchefeu-Meynier (1992).

68 *Many teratologists have linked the deformity to the myth.* Aetiological explanations of myths have the delightful property that they are more or less unfalsifiable, but that hasn't stopped many from proposing them. For arguments that the forms of the various monsters of Greek mythology (the Cyclops, the Hydra, Typhon, the Harpies etc.) are derived from foetuses that show various abnormalities see Schatz (1901). For the claim that the Cyclopes were inspired by the Pleistocene remains of dwarf elephant skulls that have been found on Sicily see Mayor (2000). Far more sophisticated discussions of deformity in ancient Greece and Egypt by a classicist can be found in Véronique Dasen's numerous papers. The relationships and symbolic meanings of the Cyclopes are discussed by the pre-eminent scholar of comparative mythology G.S. Kirk (1974) pp.85, 207.

69 *Homer to Vrolik.* Pliny the Elder *Natural history: a selection.* (trans. J.F. Healy. 1991. Penguin, Harmondsworth, UK). For an early and important outline of the origin of the Plinian races and their fate in medieval literatures see Wittkower (1942). For much further detail see the very readable Friedman (1981) as well as Kappler (1980), Williams (1996) and Daston and Park (1998).

71 *The first illustration of a cyclopic child.* Liceti (1634) also identifies an additional two eyes on the back of the Firme child's head, but these must be fictitious (he never saw the child himself, but worked from a sketch). The original sketch of the Janus-headed twins with cyclopia is now lost. Some, but not all, features of the presentation and subject in Liceti's engraving are consistent with it having been derived from an original by Leonardo. No other sketches of teratologies by him are known (Martin Kemp, pers. comm.). The report of a Janus-headed twin with cyclopia is given in Abbott and Kaufmann (1916).

72 *Looking at his bottled babies.* For a historical account of causal theories of cyclopia in the nineteenth and twentieth centuries see Adelmann (1936). For the modern definition of holoprosencephaly see Cohen (2001); for the incidence of the disorder see Muenke and Beachy, (2000); for a review of the teratology, Cohen and Shiota (2002); and for the role of veratrum in lamb cyclopia, Incardona et al. (1998).

74 *Most cases of cyclopia.* Holoprosencephaly (HPE) consists of at least seven distinct inherited syndromes (HPE1 to HPE7). HPE3 (**142945**) is due to heterozygosity for mutations or deletions in sonic hedgehog (**600725**) (Roessler et al. 1996). For the spectrum of Shh mutations see Nanni et al.

(1999, 2001). For the Shh knockout mouse see Chiang et al. (1996); for a general revew of the other HPE genes see Muenke and Beachy (2000).

76 *An embryo's face.* See Hu and Helms (1999) for an elegant study of the role of Shh in craniofacial formation and Ditto the pig. Gli3 (**165240**) mutations cause Greig's cephalopolysyndactyly (**175700**) which is characterised by hypertelorism among other things. Hypertelorism with nasal bifurcations is characteristic of craniofrontonasal syndrome (CFNS) (**304110**). The causal gene underlying has not yet been identified.

78 *Among the disorders that appear.* For a history of sirenomelia and siren-like iconography see Gruber (1955); for a more recent review see Valenzano et al. 1999). For the CYP26A1 (**602239**) deletion mice see Sakai et al. (2001) and Abu-Abed et al. (2001). For pigs without eyeballs see Hale (1933); for isotretinoin in humans see Lammer et al. (1985); and for a review of retinoic acid function and gradients in the embryo see Maden (1999).

83 *The consequences of cells.* The claim that the lumps on the necks of Pans are supernumerary auricles is made by Sutton (1890); Cockayne (1933) discusses goats. See Boardman (1997a) pp.36–7 for the Hellenistic Pans bearing supernumerary auricles. See Boardman (1997a; b) for a technical synopsis of the history of Pan in Greek and Roman art and a charming essay on his iconography by the famous scholar of Greek art. The girl with four auricles is discussed by Birkett (1858); a more general discussion of supernumerary auricles is given by Bateson (1894) pp.177–80 and Cockayne (1933) pp.339–41.

86 *Homeosis was first identified.* Bateson's (1894) purpose was not, as now, to use homeotic variation as a means for studying development. He was instead struggling towards a theory of inheritance; that is, attempting to fill the gap left by Darwin's account of evolution. He failed, but he was among the first to retrieve Gregor Mendel's experiments on peas from the fathomless obscurity of the *Verhandlungen des naturforschenden Vereines in Brünn*. As such, he is recognised today as one of the fathers of modern genetics.

87 *Over the last eighty-odd years.* See Lawrence (1992) for an account of the homeotic genes in *Drosophila*. Strictly speaking these genes specify parasegments, divisions in the embryo that contribute to, but are out of register with, the segments visible in the larva. The seminal paper on the homeotic genes is Lewis (1978). In 1995, Ed Lewis shared the Nobel Prize with two other Drosophilists, Christiane Nüsslein-Volhard and Eric Wieschaus.

91 *Extra ribs have always caused trouble.* See Sir Thomas Browne's *Pseudodoxia epidemica, or, enquiries into very many recieved tenents and commonly presumed truths* (1646) in *Works* (1904) volume III chapter II pp.5–8. Estimates on the variation in rib number come from Bornstein and Peterson's 1966 study of 1239 skeletons. They found that 9 per cent of their skeletons had thirteen pairs of ribs. Of this fraction, just under 1 per cent were due to the seventh cervical vertebra gaining ribs, about 5 per cent were due to the first lumbar vertebra gaining ribs and 3 per cent were due to a simple increase in the total number of vertebrae, that is, were not due to a homeotic transformation of vertebral type. Other studies, reviewed in Galis (1999), put the incidence of seventh cervical vertebrae ribs somewhat lower, at around 0.2 per cent. Cervical ribs can cause 'thoracic outlet syndrome', a compression of the nerves and blood vessels of the neck. Galis (1999) also addresses the fascinating question of why all mammals have just seven cervical vertebrae where the numbers of other vertebrae vary greatly among species. She argues that mutations that cause cervical ribs may be far more pathological than is generally appreciated and so under strong stabilising selection.

92 *It is no surprise, then, that the identity of each vertebra.* This account of C7 specification is based on the results of deletions for particular Hox genes in mice. It is probably incomplete and some of the AND statements should be OR (since partial transformations are common, suggesting that other Hox genes can compensate) or else couched in more quantitative terms – but it is a reasonable start. Deletions in the following genes cause C7->T1 transformations: Hoxa4 (Horan et al. 1994); Hoxa5 (Jeanotte et al. 1993); Hoxb5 (Rancourt et al. 1995); Hoxa6 (Kostic and Capecchi, 1994); Hoxb6 (Rancourt et al. 1995). Other genes, such as Hoxc4, may be affect this vertebra as well, but results disagree (Boulet and Capecchi 1996; Saegusa et al. 1996; Horan et al. 1995a; b)

93 *Distinguishing one vertebra from another.* These disorders are caused by deletions in the following genes: anteriorised limbs: Hox5b (Rancourt et al. 1995); partly missing hindbrains: Hoxa1 (Lufkin et al. 1991; Mark et al. 1993; Carpenter et al. 1993); hernias: Hoxd4 (Ramirez-Solis et al. 1993); no thymus: Hoxa3 (Chisaka and Capecchi 1991); unable to walk: Hoxc8 (Le Mouellic et al. 1992).

94 *The Hox gene calculator.* For Hox gene expression in human embryos see Vielle-Grosjean et al. (1997). The embryos used in this study were 'collected with full ethical permission'. For a discussion of what this means and legislation of such studies in various countries see Burn and Strachan (1995).

95 *Writing of the 'calculator of fate'.* For supernumerary eyelashes or distichiasis (**126300**) see Cockayne (1933) p.330 who notes, incidentally, that hedgehogs normally have two rows of eyelashes. For the seven-hearted chicken see Taussig (1988) following a 1904 report by the pathologist Verocay who happened to be staying at the inn and who managed to secure the viscera, but not the rest of the chicken, for study. Isidore Geoffroy Saint-Hilaire (1832–37) volume 1 pp.723–9 discusses various putative cases of heart duplications in humans but can come up with only one possibly authentic example, an early-eighteenth-century case of a grossly deformed infant. See Lickert et al. (2002) for extra hearts in β-*catenin*-conditional null mutant mice.

96 *And then there is Disorganisation.* Disorganisation (**223200**). The *Disorganisation* mutation was first studied by Hummel (1958, 1959), then by Crosby et al. (1992). The possibility of a human homologue of Disorganisation was mooted by Winter and Donnai (1989) and Donnai and Winter (1989), who proposed its existence to explain children whose malformations had been previously attributed to a miscellany of other cause. Disorganisation is caused by a dominant mutation on mouse chromosome 14.

97 *The power of the homeotic genes.* The evolution of snake limblessness is discussed by Cohn and Tickle (1999). This explanation for the loss of forelimbs in snakes does not account for the loss of the hind-limbs – which is due to a failure of the limb-buds to grow. Many studies have shown comparable changes in Hox gene expression patterns, particularly in arthropods. And some very recent studies have actually demonstrated that mutations in Hox genes are directly responsible for evolutionary changes in morphology (as a change in Hox gene expression, the snake example implies a mutation in some upstream regulatory factor). Mutations in Emx2 (**600035**), a human homologue of ems, is responsible for schizencephaly (**269160**). For a discussion of ems and other conserved brain genes see Reichert and Simeone (2001). Mutations in Pax6 (**607108**), a human homologue of eyeless, mutations cause aniridia (**106210**) (Ton et al. 1991). For a review of the conservation of eyeless/Pax6 in eye evolution see Gehring and Ikeo (1999).

99 *In the cyclical way of intellectual fashion.* Geoffroy's major ideas on what we now call homology can be found in his *Philosophie anatomique. Des organes respiratoires sous de la détermination et de l'identité de leurs pièces osseuses* (1818) and *Considérations générales. Sur la vertèbre* (1822) which have been collected by Le Guyader (1998) who also discusses the dispute with Cuvier as does Appel (1987). The revival of the dorso-ventral inversion hypothesis is due to, among others, De Robertis and Sasai (1996); for a sceptical update see Gerhart (2000).

CHAPTER IV: CLEPPIES

105 For the mark of Cain see Friedman (1981) pp.87–107. The football coach (p.106) was Glenn Hoddle. He was sacked (*The Times*, London, 1 February 1999).

107 *As recently as 1900.* The story of the Cleppies is told by the British geneticist Karl Pearson (1908) in one of the first studies of a 'lobster-claw' family. Most British historians, Macaulay among them, accept that the Wigtown martyrs existed, but some such as Irving (1862) have noted that there are no eyewitness accounts and doubts that the whole thing happened, the graves notwithstanding. The best account is Fraser (1877). Irving and Fraser also note that another legend has another officer, the Provost, saying to maid Wilson, 'Hech, my hearty! tak anither drink,' only to find himself evermore afflicted with an unappeasable thirst. The uncertain nature of the story is made even clearer by the reference to 'Good King Charlie' – Charles II – who at the time of the execution, 11 May 1685, had been dead for two months. Historians generally blame his successor, the Roman Catholic James II, for unleashing the army on the Scottish Lowlands. The etymology of 'clep' is also confusing. W.A. Craigie, in his *Dictionary of the older Scottish tongue* (1931), gives 'clep' as 'call', but Pearson's story suggests that 'clepped' also means to have a limb deformity. The clepped families themselves are described by Pearson (1908), McMullen and Pearson (1913) and Lewis and Embleton (1908). Pearson's papers are of particular historical interest for he uses them to advance the agenda of the biometricians against the Mendelians by showing that this apparently dominant gene does not segregate in Mendelian ratios. While his campaign against Mendelianism proved futile, he was partly right about this trait: it looks as though at least one ectro-dactyly allele is over-represented in male progeny, an apparent case of mei-otic drive, the only one known in a human pathology (Jarvik et al. 1994).

109 *The fragments of myth.* Euterpe Bazopoulou-Kyrkanidou (1997) argues persuasively that Hephaestus' lameness was usually represented as club-feet. Aterman (1999) proposes that Hephaestus' deformity is related to the achon-droplasia of the Egyptian deity Ptah – on which more in Chapter V – and, later, to arsenic neuritis, an acquired disease associated with smiths. These points of view are not necessarily inconsistent as the iconography clearly evolved over time. For the origins of the story of the Ostrich-Footed Wadoma see Gelfland et al. (1974); Roberts (1974); articles resurrecting the myth (e.g. Barrett and McCann 1980) and genetic investigations (Farrell 1984, Viljoen and Beighton 1984). Limb defects are second only to congeni-tal heart defects in frequency (Bamshad et al. 1999).

111 *One of the strange things about limbs.* Pearson (1908) and Lewis and Embleton (1908) recount the manual dexterity of 'lobster-claw' families. See Hermann Unthan's (1935) memoirs for an edifying account of armlessness. The goat is described by Slijper (1942).

113 *What induces a limb-bud to grow out into space?* The original description of the apical ectodermal ridge (or AER) and its experimental removal is described by Saunders (1948). Acheiropody (**200500**) is described by Freire-Maia (1975, 1981).

115 *The apical ectodermal ridge is the sculptor of the limb.* 'Lobster-Claw syn-drome' and 'ectrodactyly' are both now less commonly used than 'split-hand-split-foot-malformation' syndrome (SHFM). The disorder occurs in 1 in 18,000 newborns; inheritance is usually dominant. There are at least three distinct SHFM loci in humans: SHFM1 at 7q21.3-q22.1 (**183600**); SHFM2 at Xq26 (**313350**), SHFM3 at 10q24 (**600095**), and we can add a fourth, ectrodactyly, ectodermal dysplasia and cleft lip syndrome (EEC) at 3q27 (**129900**) (Celli et al. 1999). There are many other related syndromes besides. Celli et al. (1999) identify the EEC gene as p63, a close relative of the tumor suppressor gene, p53; Yang et al. (1999) and Mills et al. (1999) study its function in mice. Another ectrodactyly gene in mice, Dactylplasia, encodes an F-box/WD40 family protein thought to be involved in protein destruction, and although the human homologue of this gene maps near to SHFM3, it has not yet been shown to be causally involved (Crackower et al. 1998; Sidow et al. 1999). The same is true for two distal-less related genes, DLX5 and DLX6, thought to be responsible for SHFM1 (Merlo et al. 2002). Both p63 and Dactylplasia are involved in the maintenance of the AER; among their many other skin defects, p63-homozyous mice have no limbs at all.

116 *Action at a distance in the embryo.* Developmental biologists will notice that the account given here, which focuses on the AER's role in promoting the growth of the limb-bud, is not that given in textbooks. There is no mention of how the AER patters the proximo-distal axis of the limb-bud via the 'Progress-Zone clock' (Wolpert 1971). This is because a pair of recent papers (Sun et al. 2002; Dudley et al. 2002) have convincingly shown that the Progress-Zone clock model is wrong. This is fascinating, but a bit upsetting, since it seems to throw the question of proximo-distal patterning open again. Niswander et al. (1993) describe how beads soaked in FGF can replace the apical ectodermal ridge. Sun et al. (2002) also give the most recent account of what is now a plethora of engineered FGF mutations in mice which have shed light on how they work.

116 *Ridge FGFs not only keep mesodermal cells proliferating.* The role of FGFs in regulating cell death is shown by Dudley et al. (2002). See Zou and Niswander (1996) for the role of cell death in eliminating inter-digital webbing in chickens but not ducks. Webbing in humans, more precisely syndactyly, is sometimes the result of an excess of FGF signalling caused by gain-of-function mutations in the FGF receptor, FGFR2, as in Apert syndrome (**101200**; **176943**) (Wilkie et al. 1995).

118 *This account of the making of our limbs.* The role of thalidomide in phocomelia was first reported by McBride (1961) and Lenz (1962). Phocomelia appears in Roberts's syndrome (**268300**) and SC Phocomelia syndrome A (**269000**), which may be the same disorder and are known as 'pseudothalidomide' syndromes; the genetic basis of neither is known. Goya's sketch of a phocomelic infant is in the Louvre; Vrolik (1844–49) depicts Pepin; a brief account of his life is given in Gould and Pyle (1897) p.263.

120 *How does thalidomide have its devastating effects?* Stephens et al. (2000) reviews some of the voluminous literature on thalidomide. He firmly discounts recent sensationalistic claims that thalidomide-induced phocomelics (who are now in their late thirties) are giving birth to phocomelic children – which, if true, would imply the existence of some form of Lamarkian inheritance. In principle, however, thalidomide might be a general mutagen causing high frequencies of all sorts of genetic disorders in second-generation infants. Exhaustive studies have failed to show that this is so. Until recently, the best account of the action of thalidomide on limb formation was given by Tabin (1998). His explanation, which he convincingly defended against others (Neubert et al. 1999; Tabin 1999), rested on the idea that thalidomide causes a disassociation between proliferation and proximal-distal specification of limb-buds. In other words, it was couched in terms of the 'Progress Zone' model of limb specification. With the demise of that model (Sun et al. 2002; Dudley et al. 2002) the specificity of thalidomide becomes a little more difficult to explain but still probably depends on the abnormal inhibition of proliferation in particular populations of bone-precursors. It is striking that FGF8-conditional limb mutants in mice have phocomelia (Lewandoski et al. 2000; Moon and Capecchi 2000).

121 *Metric, with its base* 10 units. Until recently it was held that *all* modern vertebrates (living or not) have no more than five digits (Shubin et al. 1997). True, some creatures such as pandas and moles *appeared* to have six, but they could be dismissed as not being true fingers, but rather modified wrist bones (the radial sesamoid in pandas and falciform bone in moles). Polydactyly can, however, evolve in flippers such as the paddles of the icthyosaur, *Opthalmosaurus*, which appear to conceal eight digits (Hinchliffe and Johnson 1980 p.56), and those of the *vaquita* dolphin, which have six (Ortega-Ortiz and Villa-Ramirez 2000). Alberch (1986) discusses polydactylous dogs; Lloyd (1986) does so for cats; and Wright (1935) for guinea pigs. Galis (2001) reviews the question of why, despite the frequency of poly-dactylous mutations, so few species exist with more than five digits per limb. Polydactly in humans (**603596**) and many other entries). Frequencies and kinds of polydactyly from Flatt (1994); in the Ruhe family (Glass 1947); in the Scipion family (Manoiloff 1931).

122 *If the apical ectodermal ridge.* For the discovery of the zone of polarising activity see Saunders and Gasseling (1968); for its interpretation see Tickle et al. (1975). Sonic hedgehog (**600725**) was first identified as the gene encoding the morphogen by Riddle et al. (1993). Since then, some (Yang et al. 1997) have argued that it is not the morphogen since it does not form a gradient in the limb. More recent evidence suggests that it does (Zeng et al. 2001).

126 *This catalogue of mutations.* Many polydactyl genes have been identified in mice and humans, and many are transcription factors. For example, mutations in Gli3 (**165240**), a zinc-finger transcription factor, cause Greig's cephalopolysyndactyly (**175700**), Pallister-Hall syndrome (**146150**) and postaxial polydactyly (**174200; 174700**). See Manouvrier-Hanu et al. (1999) for a brief review of others. On-line Mendelian Inheritance in Man (August 2002) lists ninety-seven disorders with polydactyly in the clinical synopsis. How many of these are genuinely different is an interesting question, but the suggestion is certainly that more than ten genes are involved in correctly determining Shh activity. The Shh regulatory mutation causes extra thumbs and index fingers, more broadly, preaxial polydactyly (**190605; 174500**). The genetics are complicated. Zguricas et al. (1999) mapped the mutations, deletions and translocations to 7q36, close to the Shh gene. Clark et al. (2001) showed that these mutations deleted a portion of Lmbr1 (**605522**), a gene near sonic hedgehog, and inferred that Lmbr1 was causal. Lettice et al. (2002), whose interpretation I follow here, provide evidence that 7q36 polydactyly mutations are due to deletions of sonic hedgehog cis-acting regulatory elements that lie within a Lmbr1 intron rather than Lmbr1 itself. Achieropody (**200500**), which also maps to 7q36, has a similarly complex history. Achieropody mutations also delete Lmbr1 and, again, this gene was thought to be causal (Ianakiev et al. 2001; Clark et al. 2001), but is also probably due to a Shh regulatory mutation – though the jury is still out (Lettice et al. 2002). Certainly, the similarity of acheiropody to the pawless limbs of Shh-null mice is striking (Chiang et al. 1996; Chiang et al. 2001).

127 *Around day 32 after conception*. For the gross development of the human limb see Hinchliffe and Johnson (1980) p.75 and Ferretti and Tickle (1997). The condensations are described by Shubin and Alberch (1986). For a Hoxa13 mutation in man causing hand-foot-genital syndrome, (**142959**; **140000**) see Mortlock and Innes (1997). Mouse models: Fromental-Ramain et al. (1996) and Mortlock et al. (1996). For a Hoxa11 mutation causing radioulnar systosis (**142958**; **605432**) see Thompson and Nguyen (2000); Hoxd13 (gain of function) (**142989**), Muragaki et al. (1996); Hoxd cluster deletion, Del Campo et al. (1999). For the most comprehensive attempt at determining what the Hox genes are doing in the limb see Zákány et al. (1997), who report the effects of knocking out a variety of Hoxa and Hoxd genes in combination.

128 *Limbs are not the only appendages*. For Hox mutations that cause both limb and genital defects in humans see the preceeding note and Kondo et al. (1997). Penis size and foot length (Siminoski and Bain, 1993). For the roles of FGFs and sonic hedgehog in genitals see Perriton et al. (2002) and Haraguchi et al. (2000).

131 *The result is rather puzzling*. On the homology of lobe-finned fish fins to tetrapod limbs see Shubin et al. (1997) for a review. Sordino et al. (1995) describe Hox gene expression patterns in zebrafish compared to tetrapods. Cohn and Bright (2000) review zebrafish fin development. Dollé et al. (1993) report the Hox d13 knockout in mice; Zákány et al. (1997) argue for the successive accretion of Hox genes in evolution. The first edition of Darwin's *The variation of animals and vegetables under domestication* was published in 1868; Gegenbauer's critique in 1880. The Darwin quote is from the second (1882) edition of *The variation* pp.457–8 where he retreats. Coates and Clack (1990) describe *Acanthostega*'s limbs.

CHAPTER V: FLESH OF MY FLESH, BONE OF MY BONE
137 *Around 1896, a Chinese sailor named*. Arnold and his descendants are

described by Jackson (1951) and Ramesar et al. (1996). Their disorder was cleidocranial dyplasia (**119600**) caused by a dominant haploinsufficient mutation in the osteoblast transcription factor CBFA1 gene (**600211**). See Komori et al. (1997), Mundlos et al. (1997) and Mundlos (1999) for the identification of the mutation, its function in mice, and a review of the disorder. One of the minor puzzles of this disorder is the absence of apparent homozygous infants in South Africa. With so many carriers living in a small community, two carriers must surely have occasionally married. If the mutation works in humans as it does in mice (and there is every indication it does), one quarter of the children from such a marriage would be completely boneless and stillborn (and half would be partly boneless and one quarter would be normal).

138 *Perhaps because they are the last of our remains to dissipate.* For a more general review of bone growth see Olsen et al. (2000). For a review of the bone morphogenetic proteins see Cohen (2002). The emphasis placed here on the role of BMPs in making condensations is a little controversial; the evidence from mouse mutations tends to support a role for BMPs in patterning rather than osteoblast and chondrocyte differentiation or condensation formation (Wagner and Karsenty 2001). I suspect that this is due to redundancy among BMPs.

140 *By one of those quirks of genetic history.* Sclerosteosis (**269500**) is caused by recessive mutations in sclerostin, a secreted protein (**605740**). Until recently it was thought that the South African families (who are all Afrikaaners), a family in Bahia, Brazil, and Dutch families with a similar disorder called Van Buchem's disease or hyperostosis corticalis generalisata (**239100**) were all related, however remotely. However, the Afrikaaner, Bahia and Dutch families have now all been shown to carry different mutations in or near the SOST gene, so they cannot be related, and the presence of a similar disorder in all three populations is merely a coincidence (Brunkow et al. 2001; Balemans et al. 2001). The fused-finger disorder is proximal symphalangism syndrome (**185800**) caused by dominant haploinsufficient mutations in noggin (**602991**) (Gong et al. 1999). For null noggin mutations in mice see Brunet et al. (1998).

141 *The disorder is known as.* Fibrodysplasia ossificans progressiva or FOP (**135100**), caused by dominant mutations in an unknown gene. In 2001, a French group reported noggin mutations in FOP patients (Sémonin et al. 2001), but this could not be replicated (Cohen 2002). For Harry Eastlack's clinical history see Worden (2002) pp.185–6. For a nice essay about FOP, the people afflicted by it, and the search for its cure, see Maeder (1998).

144 *A newly born infant has.* Baker (1974) notes that the brain case of most adults is about 5 millimetres; of Australids it can be 10 millimetres; Kohn (1995) briefly discusses the head-beating ritual.

144 *What makes bones grow to the size that they do?* For an account of Victor Twitty's experiments see Twitty and Schwind (1931) and Twitty (1966).

MUTANTS

These experiments were carried out in the laboratory of the great developmental biologist Ross Harrison at Yale who initiated them (Harrison 1924).
See Brockes (1998) for a review of salamander limb regeneration.

147 *The man whose name*. For a biography of Mengele see Posner and Ware
(1986).

148 *Among those spared*. The account is partly based on Elizabeth Ovitz's
memoir (Moskovitz 1987) which is also the source of quotations. A careful
study of the family and their experience at Auschwitz-Birkenau (Koren and
Negev 2003) has, however, shown numerous inaccuracies in the memoir.
Detailed accounts of the medical experiments on human subjects carried out
in the Third Reich can be found in Lifton (1986).

154 *Pseudoachondroplasia – the disorder that afflicted*. Pseudoachondroplasia
(**177170**) is caused by dominant gain-of-function mutations in the cartilage
oligomeric matrix protein gene (**600310**) (Briggs and Chapman 2002). The
diagnosis of the Ovitzes as having this disorder rather than achondroplasia
(as is often stated) is given in Koren and Negev (2003) and is consistent with
their attractive facial features.

154 *Achondroplasia is caused*. Achondroplasia (**100800**) is caused by dominant
gain-of-function mutations in the fibroblast growth factor receptor 3 gene
(**134934**) Rousseau et al. (1994); Bonaventure et al. (1996). For a history of the
iconography of dwarfism see Dasen (1993; 1994) and Aterman (1999).

156 *If an excess of FGF signalling*. For the role that FGFs play in limb growth
see Naski et al. (1996; 1998) and Chen et al. (2001). Colvin et al. (1996) study
the FGFR3 knockout mouse; De Luca and Baron (1999) review FGFR3
function.

156 *Achondroplasia is a relatively mild disorder*. Thanatophoric dysplasia
(**187600**) is caused by severe dominant gain-of-function mutations in
FGFR3 (Rousseau et al. 1995; Tavormina et al. 1995). Oostra et al. (1998b)
describes the Vrolik skeletal dysplasia specimens.

157 *FGF must be only one molecule among many*. The authoritative review of
overgrowth syndromes is Cohen (1989). Myostatin (**601788**) McPherron et
al. (1997) for the mouse mutation; McPherron and Lee (1997) for cattle. The
original myostatin mutation occurred naturally on a Flemish farm and so
Belgian Blue meat is made ubiquitously into hamburger. Had the same animal been engineered by Monsanto it would have been surely rejected by a
public ever suspicious of 'genetically modified foods'.

158 *Mutations that disable bone collagens*. There are several types of osteogenesis imperfecta. The most common type that is not lethal at birth is
osteogenesis imperfecta type 1 (**166200**) caused by dominant hapoloinsufficient
or gain-of-function mutations in the collagen 1A2 or collagen 1A1 genes
(**120150**; **120160**) (Olsen et al. 2000).

160 *Even once our growth plates*. See Blair (1998) and Günther and Schinke
(2000) for reviews on osteoclast function and specification.

160 *There are many ways to upset the balance.* Malignant autosomal recessive osteopetrosis (**259700**) is caused by recessive mutations in genes that encode part of the vacuolar proton pump need for hydrochloric acid transport (Kornak et al. 2000).

161 *The shortness of Henri de Toulouse-Lautrec.* The biographical material and anecdotes come largely from Frey's (1994) authoritative biography. See Lazner et al. (1999) for the relationship between osteopetrosis and osteoporosis. Maroteaux and Lamy (1965) diagnosed Lautrec with pycknodysostosis and review older diagnoses; see Frey (1995 a; b) and Maroteaux (1995) for the exchange concerning his malady. Pycnodysostosis (**265800**) is caused by recessive mutations in the Cathepsin K gene (**601105**) (Gelb et al. 1996).

CHAPTER VI: THE WAR WITH THE CRANES

169 *From the walls of the Prado.* For the iconography of dwarfing see Tietze-Conrat (1957) and Emery (1996). Tanner (1981) pp.120–1 discusses Geoffroy and Buffon on dwarfs.

170 *Were all the court dwarfs unhappy.* See Boruwlaski's memoirs (1792) and Heron (1986) for a modern account of his life.

175 *At the base of our brains.* See Laycock and Wise (1996) for the regulation of growth by the hypothalamus-pituitary pathway. Primary growth-hormone deficiency (**262400**) is caused by recessive mutations in the growth-hormone gene (**139250**). There is an enormous literature on this group of syndromes; see López-Bermejo et al. (2000) for a brief review.

176 *Joseph Boruwlaski has all the signatures.* The Ecuadorean dwarfs have Laron- or growth-hormone resistance syndrome (**262500**) caused by recessive mutations in the growth-hormone receptor gene (**600946**); see Rosenfeld et al. (1994) and Rosenbloom and Guevara-Aguirre (1998).

177 *In 1782 Joseph Boruwlaski met.* Frankcom and Musgrave (1976) write about Patrick Cotter; Bondeson (1997) about Charles Byrne; Thompson (1930, 1996) tells of both men as well as other famous eighteenth-century giants.

179 *Charles Byrne had a pituitary tumor.* See Keith (1911) for original diagnosis.

179 *An old photograph shows a triptych of skeletons.* See Schnitzer (1888) (Emin Pasha's given name) for an account of how he obtained the skeletons. See Schweinfurth (1878) for an account of meeting Akadimoo. The Homeric quote is from *The Iliad* (1950 trans. E.V. Rieu, Penguin Books, Harmondsworth, UK). Tyson (1699, 1966); Schweinfurth (1878); de Quatrefages (1895); Cavalli-Sforza (1986) all give accounts of Greek and Roman writings on pygmies, but the authoritative work on pygmies in antiquity is Dasen (1993). See Addison (1721) for his verses on pygmies.

182 *Addison's poem.* See Tyson (1699, 1966) for his dissection of a 'pygmie'; de Quatrefages (1895) says that the word 'Aka' can be found inscribed on the frescos of a fifth-century Egyptian tomb beneath a depiction of a dancing pygmy; Cavalli-Sforza (1986) repeats the story. This would be remarkable if true, but sadly Véronique Dasen assures me it is not.

183 *The French anthropologist Armand de Quatrefages*. See Schebesta (1952) and Weber (1995–99) for the history of the negritos; Diamond (1991) discusses theories about pygmy smallness. For the most recent study on the genetic relationships of Andaman Island negritos see Thangaraj et al. (2003).

185 *The diagnosis of achondroplasia*. The Attic vase is just one of many examples of Greek pygmy iconography given by Dasen (1993). Gates (1961) asserted that pygmies have achondroplasia.

185 *That pygmy proportions*. The genetics of pygmy smallness are obscure, but the evidence seems to exclude a single locus with substantial dominance. See Shea and Bailey (1986), Shea and Gomez (1988) and Shea (1989) for an analysis of pygmy proportions.

187 *The geographers, entranced by their acquisition*. The history of Chair-Allah and Thibaut is given in de Quatrefages (1895) and Schweinfurth (1878); their growth curve is given in Cavalli-Sforza (1986) p.366.

188 *A newborn infant grows*. See Tanner (1990) p.12 for the pubertal growth spurt and Tanner (1981) pp.104–5 for the history of its study; see Bogin (1999) for pubertal spurts in other primates.

189 *The pubertal spurt is driven*. There is some disagreement as to whether pygmies show low IGF-1 serum titres or whether they have a less effective IGF-1 receptor relative to taller people. In any event, what we know about short stature in pygmies is based on endocrinological studies (Merimee et al. 1981; 1987; Geffner et al. 1995). The following note also bears on the interpretation of these data.

189 *The proof of this is the mini-mouse*. The account of the relationship between IGF and GH given here differs from the 'somatomedin hypothesis' given in most textbooks. Recent experiments have suggested that: (1) IGF-1's primary role is as a paracrine (short-range) growth factor rather than an exocrine hormone; (2) that liver IGF-1 is responsible for most serum IGF, and that it contributes little to post-natal growth; (3) that IGF-1's effects on growth are therefore to a considerable degree – though not entirely – independent of GH's. See Le Roith et al. (2001) for a review of these matters, and Lupu et al. (2001) for an account of the GHR; IGF-1 double knockout mouse.

190 *Schweinfurth's discovery set off a global hunt*. See Haliburton (1891; 1894) for 'pygmies' in Spain, the Atlas Mountains and Switzerland. See Johanson and Edgar (1996) for the stature of fossil hominids and Bogin (1999a) p.3 for the 'Maya in Disneyland'.

192 *It is even possible that the most recent*. Exploration accounts can be found in Kingdon-Ward (1924; 1937) and many other books by the same author. I first learned about the Burmese 'pygmies' from Prof. Harry Saing, who saw them while travelling in Kachin State in 1964. The Burmese government reports are given in Mya tu et al. (1962, 1966).

194 *It is not a pretty word*. For a general review of iodine deficiency diseases and cretinism see Delange and Hetzel (2000). Several reports from western China show a form of myxedmatous cretinism that has many of the features seen in the Taron. Various co-factors have been suggested for the extreme form of cretinism seen in the Congo such as selenium deficiency or a fondness for eating goitrogenic plants.

195 *Cretinism is a global scourge*. See Merke (1993) for Napoleon's investigations into Swiss cretins and the iconography of cretinism in Aosta Cathedral. See Delange and Hetzel (2000) for the discovery of the thyroid and Laycock and Wise (1996) pp.203–40 for how it works. The hormone called here 'thyroxine' is, more formally, two hormones, tri and tetraiodothyronine or T_3 and T_4. Williams et al. (1998) discusses the cellular role of thyroxines on the growth plate. Among the many genes with known mutations causing a deficiency of thyroxine production or function are: thyroid peroxidase, an enzyme that is involved in thyroid hormone production (**188450**) (Abramowicz et al. 1992); thyroglobulin (**606765**), or the thyroid hormone receptor (**190160**) (Refetoff et al. 1996); as well as several other genes involved in the synthesis, transport or storage of thyroid hormone (de Vijlder et al. 1999; Vassart 2000). Mutations in all these genes tend to cause cretinism with goitre.

197 *There is also a class of mutations more vicious by far*. Among the genes with known mutations that affect the pituitary's stimuation of the thyroid are those that encode: thyroid stimulating-hormone (TSH) itself (**188540**) (Hayashizaki et al. 1989) and those that cause combined pituitary hormone deficiency (CPHD) due to mutations in transcription factor genes such as PROP-1 (**601538**) and PIT-1 (**173110**) required for the specification of somatotroph, thyrotroph and lactotroph cells (Tatsumi et al. 1992; Voss and Rosenfeld 1992; Sornson et al. 1996; Wu et al. 1998). These mutations cause cretinism without goitre and with or without dwarfism.

199 *Nearly twenty-five centuries ago*. For Aristotle on the effects of castration see his *Historia animalium* in *Complete works* volume 1 pp.981–2. For a more recent view of the same subject see Wilson and Roehrborn (1999).

203 *We think of the estrogens*. For estrogen-receptor deficiency (**133430**) in men see Smith et al. (1994). The enzyme that converts testosterone to estrogen is aromatase cytochrome P450 (**107910**) For loss-of-function mutations that cause continued growth in men see Sharpe (1998); Lee and Witchel (1997); for gain-of-function mutations that cause excess in women see Stratakis et al. (1998).

203 *Growth hormone and IGF are extremely powerful*. One of the several cancer-predisposition syndromes caused by mutations in PTEN (**601728**) is Cowden syndrome (**158350**). For the evidence that Proteus syndrome

(**176920**) is also caused – at least sometimes – by PTEN loss-of-function mutations see Zhou et al. (2000, 2001). The Ovid quote is from *Metamorphoses* (trans. A.D. Melville. 1986. Penguin Books, Harmondsworth, UK). Seward (1992) re-examines Joseph Merrick's skeleton in detail and upholds the traditional diagnosis of neurofibromatosis type 1 (**162200**). Tibbles and Cohen (1986), Cohen (1988) and Cohen (1993), however, argue for Proteus syndrome.

206 *The intimate relationship between growth and cancer.* For the relationship between IGF titres and dog body-size see Eigenmann et al. (1984, 1988); Eigenmann (1987). The association between osteosarcoma and size in dogs was detected by Tjalma (1966), in children by Fraumeni (1967). The latter result has been confirmed by three out of four studies since. See Leroi et al. (2003) for a general discussion on the causes of pediatric cancers. Jenkins (1998) reviews increased propensity of acromegalics to a variety of cancers. The causal role of IGF is reviewed by Holly et al. (1999). In his classic book on ageing, Comfort (1964) first noted that big dogs do not live as long as small dogs. See also Patronek et al. (1997) and Miller and Austad (1999). The best data on ageing rate, from the Swedish pet health insurance scheme, is given in Egenvall et al. (2000).

208 *I am fascinated by these findings.* Krzisnik et al. (1999) discuss the dwarfs of Krk who are homozygous for recessive mutations in PROP-1. Samaras and Elrick (1999) and Samaras et al. (1999) give a partisan account of the evidence for a negative association between human height and longevity. See Waaler (1984) and Power and Matthews (1997) for the general positive association between health and longevity. There is, however, some evidence from the Finnish studies of a U-shaped mortality distribution in women, possibly associated with skeleto-muscular problems in the tallest women (Läärä and Rantakallio 1996; Silventoinen et al. 1999).

208 *These results seem to tell us.* The first dwarf mice which were shown to be long-lived were Snell dwarfs (*dw*) which are deficient in a number of their pituitary gland lineages because of a mutation in PIT-1, a pituitary specific transcription factor. Ames dwarf (*df*) has the same phenotype, but has a mutation in another transcription factor, PROP-1. Both these mice are long-lived, but since they lack both somatotrophs and thyrotrophs, they fail to produce both growth hormone and thyroid-stimulating hormone, making it impossible to distinguish the effects of lacking either (Brown-Borg et al. 1996; Bartke et al. 2001a; b). However, the Little mouse (*lit*) is also long-lived (Flurkey et al. 2001). Since this dwarf mouse has a mutation in its growth-hormone releasing-hormone receptor (GHRHR), it is very likely that it is growth-hormone deficiency, or its sequelae, that cause the longevity of these strains.

209 *To be poor is to be both short and at higher risk.* See Mansholt (1987) on Dutch growth; Mackenbach (1991) on the socio-economic causes of height differences in Holland and Didde (2002) for height-related activism in the Netherlands. Also see Cavelaars et al. (2000) for a fascinating comparision of the secular increase in all European countries which shows that although all countries show a secular increase in height, they are all getting taller at more or less the same rate. There is much evidence that milk consumption is responsible for a good deal of environmental variation in height, for example, in the Japanese; Takahashi (1984); Bogin (1999a) p.268.

210 *The poverty and short stature of the north's people.* See Rosenbaum et al. (1985) and Mascie-Taylor and Boldsen (1985) for regional differences in height in England and Townsend et al. (1992) for the authoritative survey of health inequalities in Britain. Tanner (1981) p.147 discusses Chadwick and his surveys of height.

211 *It is precisely the antiquity of the positive association.* There is a huge literature on the attractions of height. Some of it reviewed by Bogin (1999a) pp.326–7. See Sandberg et al. (1994), Guyda (1998) and Root (1998) on growth-hormone therapy for short children.

CHAPTER VII: THE DESIRE AND PURSUIT OF THE WHOLE

217 *In February 1868, a Parisian.* For the journal and other relevant papers see Barbin (1980). Confusingly, Herculine Adélaïde (Alexina) refers to herself as 'Camille'.

221 *Is Alexina a woman?* Chesnet (1860) *Annales d'hygiène et de médecine légale* 2e série, XIV: 206 quoted in pp.124–8 of Barbin (1980). See Goujon (1869) for her autopsy. See Dreger (1998) for a social history of hermaphrodites.

223 *Anatomists, however, have other tastes.* Laqueur (1990) claims that Vesalius' and Galen's homologies are confirmed by modern, or rather nineteenth-century, embryology, but this is not so. Thiery and Houtzager (1997) p.51 describe the background of Vesalius' analysis of the vagina and note that Vesalius so loved his homology between the uterus and scrotum that he depicted the former with a dividing cleft comparable to the raphe of the scrotum where there is none.

225 *It was another Paduan anatomist.* Laqueur (1989) delves deeply into the history of the identity of male and female genitalia. O'Connell et al. (1998) redescribe the vestibular bulbs as the clitoris; Williamson and Nowak (1998) add further details about the discovery; Kobelt (1844) gives an earlier view.

228 *By day 28.* Descriptions and timing of embryological events are from McLachlan (1994). I have omitted a description of the internal genitalia (fallopian tubes, uterus and upper vagina, epididymus, vasa deferentia and seminal vesicles), all of which have other embryological origins and are controlled by other hormones.

230 *To develop as a female is to travel.* For an account of the discovery of the Y see Mittwoch (1973); XX[n]Y males are also often mentally retarded and sterile. The condition is known as Kleinfelter syndrome and occurs a

frequency of 1 in 1000 male births (Conner and Ferguson-Smith 1993).
231 *The search for the source of the Y's power*. The papers that describd SRY
(**480000**) in humans and mice are Sinclair et al. (1990); Gubbay et al. (1990);
McLaren (1990) gives a contemporary commentary.
233 *Perhaps SRY activates a few critical genes*. For an update of the genes
known to be regulated by SRY in the gonad see Graves (1998). For
Alexandre Jost's experiments see Jost (1946–47). The results described are
true for rabbits castrated before day 22 post coitem.
234 *Or at least it needs its Leydig cells*. The testosterone synthesis pathway can
be upset at many points. Luteinising hormone is needed for proper Leydig
cell growth. Mutations in the luteinising hormone receptor gene (**152790**)
cause Leydig cell hypoplasia and so pseudohermaphroditism (Kremer 1995
and Laue et al. 1996). Then various mutations can disrupt the testosterone
biosynthetic pathway (Besser and Thorner, 1994). Some of these cause a
group of syndromes known as the congenital adrenal hyperplasias (CAH)
(e.g. **201910**) since they affect not only testosterone synthesis but the synthesis of other steroids by the adrenal gland as well and have, accordingly, wide-
spread physiological effects. Good examples of testoterone synthesis mutations are those in the 17-β hydroxysteroid dehydrogenase gene (**605573**)
(Russell et al. 1994; Geissler et al. 1994).
235 *Such girls are, it is often said, exceptionally feminine*. Androgen insensitivity syndrome (**300068**) caused by mutations in the testosterone receptor
gene (**313700**). For the height of testosterone receptor-null people see
Quigley et al. (1992). For the identical twin flight attendants see Marshall
and Harder (1958); for the French model see Netter et al. (1958).
236 *Alexina/Abel and Marie/Germain were both isolated cases*. Montaigne, the
humanist, attributed Marie/Germain's sex change to sublimated sexual
desire; Paré, the (evidently Vesalian) surgeon, thought that the exertion of
the chase had caused Marie's genitals to fall out. Montaigne (1580; 1958) p.38;
Paré (1573; 1982) p.31. For the Dominican Republic *guevedoche* see
Imperato-McGinley et al. (1974); for the Papua New Guinea *kwolu-aatmwol*
see Imperato-McGinley et al. (1991). The mutated gene in all these cases has
either been shown, or else is presumed to be, 5-α-reductase (**264600; 607306**).
238 *When I said that the route to femininity*. For infant female pseudohermaphroditism caused by deficiency in aromatase (**107910**) see Shozu et al.
(1991); for the adult aromatase deficiency see Conte et al. (1994) and
Morishima et al.(1995). For aromatase excess (shortness, gynecomastia in
boys, large breasts in girls) due to dominant gain-of-function mutations, see
Stratakis et al. (1998).
240 *Spotted hyenas are unsympathetic creatures*. See Neaves et al. (1980);
Glickman et al. (1992); Licht et al. (1992); Holekamp et al. (1996) and Frank
(1997) for spotted hyena endocrinology, genitalia and social structure. Moles
(*Talpa*) also have a kind of female pseudohermaphroditism – although they

actually have ovotestes, so could be said to be true hermaphrodites.

242 *In The symposium*. Plato, *The symposium* pp.59–65 (trans. W. Hamilton. 1951. Penguin Books, Harmondsworth, UK). For sexual relations of the *guevedoche* and *kwolu-aatmwol* see Imperato-McGinley et al. (1991) and Herdt (1994).

CHAPTER VIII: A FRAGILE BUBBLE

247 *Our species has, since 1758*. Bendyshe (1865) gives a summary and English translation of Linnaeus' anthropological works; Pearson et al. (1913) and Broberg (1983) discuss *Homo troglodytes*. Lindroth (1983) discusses Linnaeus' intellectual roots in medieval thought.

251 *His French rival Buffon*. For an account of Geneviève see Buffon (1777) Addition à l'article qui a pour titre, Variétés dans l'espèce humaine, *Supplément à l'histoire naturelle* volume 4 pp.371–454.

253 *We are a polychrome species*. For a general review of pigmentation genetics see Sturm et al. (1998). The most common form of albinism is oculocutaneous albinism type 1 or OCA1 (**203100**), which is due to recessive mutations in the tyrosinase gene (**606933**). Albinism with grey eyes is oculocutaneous albinism type 2 or OCA2 (**203200**), due to recessive mutations in the P gene (Durham-Pierre et al. 1994; Stevens et al. 1997).

254 *In 1871, en route to his encounter with the Aka*. See Schweinfurth (1878) volume 2 pp.100–1 for his account of albinos in Africa, and Woolf and Dukepoo (1969) for albinos among the Hopi.

255 *Those children would have fascinated Buffon*. For the history of Marie Sabina see Buffon (1777) p.557. Pearson et al. (1913) and Dobson (1958). For some of the other eighteenth-century piebalds see Blanchard (1907). For Lisbey's history see Pearson (1913). Pearson's insistence on this rather forced account of the inheritance of piebaldism stems, again, from his opposition to the Mendelian theory of inheritance.

261 *Molecular devices are required*. Piebald trait, white forelock and bilateral hypopigmentation of the limbs and trunk (**172800**) is caused by dominant mutations in c-Kit (**164920**) which encodes a receptor tyrosine kinase. c-Kit's ligand is steel (*Sl*) in mouse, but no human disorder has been identified with mutations in this gene. c-Kit and its ligand are thought to help in guiding the migration of the presumptive melanocytes. The other piebald syndromes often only cause white forelock but are associated with deafness or megacolon. These are Waardenburg's syndromes types I through IV (**193500**, **193510**, **602229**), caused by dominant mutations in Pax3, Sox10 and MITF (Tassabehji et al. 1992; Watanabe et al. 1998). These genes are transcription factors needed for specification of melanocyte lineages (Goding 2000). All these syndromes manifest variably, all are caused by dominant mutations; homozygotes are probably lethal.

261 *What gives us our skin colours?* Africans differ from Europeans, and

East Asians are known to differ in the structure and density of their melanosomes (Szabó et al. 1969; Toda et al. 1972), but very little is known about the genetics – except for a hint it might have something to do with the P gene (Sturm et al. 1998). See Linnaeus (1758) pp.20–1 for his diagnosis of the human species; a translation is given by Robins (1991) p.171.

262 *For nearly half a century.* The history of race-classification in South Africa is discussed by Posel (2001). Rita Hoefling's story is told by Joseph and Godson (1988).

265 *Like growth hormone, melanotropins.* Red hair and obesity are caused by mutations in POMC (**176830**), a gene that encodes the α-MSH and ACTH precursor (Krude et al. 1998).

266 *Yet not all redheads are fat.* Red hair (**266300**) caused by recessive mutations in the MC1R gene (**155555**) (Robbins et al. 1993; Valverde et al. 1995; Smith et al. 1998; Flanagan et al. 2000; Healy et al. 2001). Besides the general plausibility arguments that I have given as to whether red hair has been selected or not (Darwin 1871, 1981 volume 2 pp.316–405; Robins 1991 pp.59–72) and is therefore properly thought of as a mutation or polymorphism, there are also elaborate statistical tests which can sometimes detect historical patterns of selection. Such tests have been applied to MC1R, but they are inconclusive (Rana et al. 1999; Harding et al. 2000).

268 *Pale, and proud of it.* For a detailed discussion of the pre-PRC history of Chinese anthropology and eugenic thought see Dikötter (1992, 1997, 1998). For a study of Ainu hairiness see Harvey and Brothwell (1969).

269 *In the collection of the Capodimonte.* For the history and iconography of the Gonsalvus family see Aldrovandi (1642); Siebold (1878); Zapperi (1995); Haupt et al. (1990) pp.92–7 and, especially, Hertel (2001).

273 *In 1826 John Crawfurd, British diplomat and naturalist.* For the history of Shwe-Maong and his family see Crawfurd (1827); Yule (1858) and Bondeson and Miles (1996).

276 *We are born with about five million hair follicles.* For a general review of hair (and feather) specification see Oro and Scott (1998). For the role of BMPs and FGFs, Jung et al. (1998) and Noramly and Morgan (1998). Reynolds et al. (1999) carry out the trans-gender transplantation experiment.

280 *The one thing that many of us.* Most of the anecdotal material here comes from Segrave (1996) – a delightful social history of balding. Male pattern balding or androgenetic alopecia (**109200**). See Cotsarelis and Millar (2001) for a general biology of the dying hair follicle, and Kuester and Happle (1984) for a review of the genetics of the androgenetic alopecia.

282 *One fact is, however, known: to go bald you need testosterone.* See Aristotle *Historia animalium* in *Collected works* pp.983–4. Hamilton (1942) recounts the experiments with testosterone. Knussmann et al. (1992) discuss the relationship between testosterone levels, virility and balding.

283 *Is there any hope for the bald?* Trotter (1928) discusses the relationship between hair growth and shaving. Sato et al. (1999) and Callahan and Oro (2001) discuss the role of sonic hedgehog in rejuvenating hair follicles; Huelsken et al. (2001) discuss β-catenin.

285 *One can still, occasionally*. The portraits of the Ambras family were first described in the modern scientific literature by the physiologist C. Th. Siebold (1878). He proposed that they were atavistic, a claim echoed by Brandt (1897), who points out that the Burmese family have the same disorder. Both men recognised that the surplus hair in the two families was lanugo (Siebold explicitly compares Petrus Gonsalvus's hair to that of a foetal orangutan), but suppose that lanugo is more 'primitive' – a conflation between phylogeny and ontogeny that is typical of German workers of the time, who were deeply influenced by Haeckel. Felgenhauer (1969) gives a summary of nineteenth-century views on hairy people. More recently, there has been a great deal of debate about just how many surplus-hair syndromes there are, and who had what (see Garcia-Cruz et al. 2002 for one point of view). I argue that Petrus Gonsalvus's and Shwe-Maong's families both have the same condition: hypertrichosis lanuginosa (**145700**), the mutant gene of which may reside on chromosome 8. The hair of at least one man with this syndrome (a Russian named Adrian Jewtichjew) has been examined microscopically and seems to have been lanugo. The most famous modern pedigree of hairy people, the Gomez family of Mexico, have another, unrelated, disorder: X-linked hypertrichosis terminalis (**145701**); Figuera et al. (1995). See this paper and Hall (1995), recent – and perhaps reasonable – claims that this latter kind of hairiness is indeed atavistic.

286 *Darwin himself knew of the Burmese hairy family*. See Darwin (1871; 1981) volume 2. p.378 for his account of sexual selection and hairiness of the Burmese family; see Darwin (1859; 1968) pp.183–4 and Darwin (1882) volume 2, pp.319–21 for the homology between skin organs, the Burmese family and the 'Hindoos of Scinde'. See Thadani (1935) for a later account of the same pedigree (the 'Bhudas') who have a syndrome called ectodermal dysplasia 1, anhydrotic or ED1 (**305100**) caused by a mutation in ectodysplasin (EDA) (Kere et al. 1996). The Mexican hairless dog's mutation is still unknown (Schnaas 1974; Goto et al. 1987) but is probably this gene or its receptor, EDAR (**224900; 604095**) (Headon and Overbeek 1999; Monreal et al. 1999). The scaleless variety of Medaka has a mutation in the EDAR gene (Kondo et al. 2001). For ectodysplasin's proposed role in establishing hair papillae see Barsh (1999). See Sharpe (2001) on the evolutionary history of the hair follicle.

288 *The use of a single molecule in the making*. For hens' teeth see the classic experiments by Kollar and Fisher (1980), a commentary by Gould (1983) pp.177–86, and recent experiments showing that chicken mandibles are BMP4-defective (Chen et al. 2000).

289 *Perhaps it is also the retrieval of an ancient signalling system*. Nipples, supernumerary or polymastia (**163700**). For a review see Cockayne (1933) pp.341–5; Japanese polymastia, Iwai (1907). I thank Alan Ashworth and Beatrice Howard for telling me about *Scaramanga*.

290 *Breasts bring us back to Linnaeus.* The ancient iconography of Artemis Ephesia is discussed by Fleischer (1984) and Linnaeus' use of it by Gertz (1948) – for the translation of which I am indebted to Lisbet Rausing. *Nosce te ipsum* – the slogan that meant so much to Linnaeus is rarely attributed to Solon, but rather (as in Plato) to the seven wise men of Protagorous who wrote it on the temple of Apollo at Delphi. *The Oxford dictionary of quotations* gives its source as 'Anonymous'.

CHAPTER IX: THE SOBER LIFE

297 *Huntington disease is one of the nastier.* Huntington disease, also Huntington's Chorea or HD (**143100**), is caused by dominant mutations in the huntingtin gene. Rubinsztein (2002) reviews the molecular basis of the pathology; Bruyn and Went (1986) review the history and spread of the disease.

298 *How can so lethal a disorder?* See Haldane (1941) pp.192–4.

300 *Were it not for ageing's pervasive effects.* Ricklefs and Finch (1995) give estimates of longevity in the absence of ageing.

302 *But it was another British scientist.* See Medawar (1952) and Williams (1957) for the seminal papers on the evolutionary theory of ageing. Rose (1991) gives an incisive historical review. Albin (1988) discusses the fecundity of women with Huntington's based on data collected by Reed and Neel (1959).

304 *In his declining years, flush with cash and fame.* Alexander Graham Bell (1918) analyses the Hyde family; Quance (1977) discusses Bell's interests in the genetics of longevity.

306 *In the 1980s the evolutionary account of ageing.* See Rose (1984) for the original experiment; Rose (1991) for a review; and Sgrò and Partridge (1999) for a more detailed analysis of a similar experiment.

308 *Since Aristotle.* Aristotle *On length and shortness of life* in *Complete Works* volume 1 p.743. See Diamond (1982) for the cost of reproduction in marsupial mice and Westendorp and Kirkwood (1998) for the cost of reproduction in British aristocrats. See Leroi (2001) for a sceptical treatment of cost of reproduction data.

309 *Is there a recipe for long life?* See Cornaro (1550, 1903) for a translation of the *Vita sobria*, and Gruman (1966) for a review of Cornaro's thought and its influence.

311 *The worst of it is that there is an element of truth.* See Finch (1990) pp.506–37 for a review of the earlier literature on caloric restriction; *ibid.* pp.20–1 for mortality rates of the Dutch during the *Hongerwinter.* See Holliday (1989) and Chapman and Partridge (1996) on reproductive costs and caloric restriction. Several experiments in flies and mice have been done to look at the effects of caloric restriction on 'whole genome expression profiles'. The best is a study on flies (Pletcher et al. 2002); the mouse studies (Lee et al. 1999) are more difficult to interpret.

313 *We term sleep a death.* See Beckman and Ames (1998) and Ames et al. (1993) for a review of the free radical theory of ageing. See Rose (1991) for SOD in gerontocratic flies. Parkes et al. (1998) for overexpression of superoxide dismutase in *Drosophila* motorneurons; Finch and Ruvkun (2001) for a general review of SOD and ageing.

316 *Our genomes contain three genes.* Familial amyotrophic lateral sclerosis or ALS1 (**105400**) is caused by dominant mutations in Cu/Zn superoxide dismutase or SOD1 (**147450**) (Rosen et al. 1993). Deleting this gene in mice seems to have little obvious phenotypic effect, although longevity does not seem to have been examined (Reaume et al. 1996). For the experiments excluding free radicals and hydrogen peroxide as a cause of ALS see Subramaniam et al. (2002); for a review, Orr (2002). For a more general discussion on the causes of ALS see Newbery and Abbott (2002). For the role of SOD1 in Down's syndrome see Epstein et al. (1987) and Reeves et al. (2001).

319 *Wrinkling is a manifestation.* Werner's syndrome (**277700**) caused by recessive mutations in RECQL2 (also known as WRN) helicase (**604611**) (Yu et al. 1989) reviewed by Martin and Oshima (2000).

319 *As we age.* For two reviews of the proposed role of cellular senesence (or Hayflick's limit) in ageing see Rose (1991) pp.126–36 and Shay and Wright (2000). Bodnar et al. (1997) show that overexpression of telomerase in human cell lines confers cellular immortality. There are some reports that the neuronal cells of mice do not undergo cellular senesence *in vitro* (Tang et al. 2001; Mathon et al. 2001). There are also strong suggestions that the proliferation of mouse cells *in vitro* is not telomere limited (Shay and Wright 2000).

321 *Mice, it seems, can get by without telomerase.* See Blasco et al. (1997), Lee et al. (1998) and Rudolph et al. (1999) for telomerase-deficient mice. One worry about these results is that laboratory mice seem to have much longer telomeres than wild mice (Weinstein and Ciszek, 2002).

322 *One way to prove the point would be to clone a human.* The original report on cloning Dolly was Wilmut et al. (1997). She died on 14 February 2003. Shiels et al. (1999) reported Dolly's short telomeres. Cloned cattle appear to have perfectly normal, indeed rather long, telomeres (Lanza et al. 2000; Betts et al. 2001). There is a controversy about the healthiness of cloned animals (Cibelli et al. 2002; Wilmut, 2002). Six generations of mice have been cloned with no sign of rapid ageing – but then, they do seem to have very long telomeres.

323 *Telomerase-mutant humans.* Hutchinson-Gilford syndrome (progeria) (**176670**) is caused by a mutation in the gene encoding Lamin A and C.

323 *In the last ten years there has been a revolution.* See Kenyon et al. (1993) for a pioneering paper in *C. elegans* ageing studies, and Leroi (2001), Finch and Ruvkun (2001) and Partridge and Gems (2002) for recent reviews.

326 *One of the first longevity genes to be identified.* Alzheimer's disease (**104300**). Late onset (AD2) is associated with particular polymorphisms in the apolipoprotein E gene (**107741**). For the relative risk of the ε4 allele see Corder et al. (1993); for its rarity in French centenarians see Schächter et al. (1994) and Charlesworth (1996).

327 *All this seems to matter less if you are black.* For the worldwide distribution of APOE alleles and discussion of relative risk of Alzheimer's among ethnic groups see Fullerton et al. (2000). There are two ideas why Africans may not feel the deleterious effects of the ε4 allele. First, haplotype analysis shows that their ε4 alleles are somewhat different from those in European populations. Perhaps it simply lacks the pathogenic effect. Second, perhaps it has exactly the same effect, but Africans have, at high frequency, a variant at another locus that protects them against ε4. There is no reason to favour one idea over the other. For African APOE allele frequencies see Zekraoui et al. (1997).

327 *In Europeans, at least, the genetics of Alzheimer's provide.* The early onset Alzheimer's genes are: AD1, εAPP (**104760**); AD3, Presenilin 1 (**104311**) and AD4, Presenilin 2 (**600759**) (Charlesworth 1996).

327 *These kinds of findings are only the beginning.* Heijmans et al. (2000) review the state of the centenarian gene hunt.

329 *In 1994 a remarkable thing happened.* Much of the discussion on late-life mortality trends is based on Wilmoth (2000) and Wilmoth et al. (2000).

CHAPTER X: ANTHROPOMETAMORPHOSIS

335 *The authors of books.* Steve Jones, in the concluding chapter of his *The language of the genes* (1993) HarperCollins, London gives a classic utopian account of humanity's future. Mark Ridley, in the concluding chapter of his *Mendel's demon* (2000) Weidenfeld and Nicolson, London suggests the wacky, but interesting, idea that we might evolve huge genomes and fantastically complex life-cycles. For the ethical views of some of the less inhibited scientists see the writings of Richard Dawkins and the late William Hamilton; for the opposition see the *New York Review of Books* (New York) and the *Sunday Times* (London).

337 *Race has long been under siege.* Steve Jones gives a good, if dated, account of these issues in *The language of the genes*. More recently, see Barbujani et al. (1997) and Rosenburg et al. (2002) for studies based on microsatellite loci; and Stephens et al. (2001) for single nucleotide polymorphisms.

339 *The variants are known as AIMS.* For an account of the search for AIMS see Collins-Schramm et al. (2002) and Shriver et al. (2003). For an account of the molecular genetics of FY (also known as Duffy) see Li et al (1997).

340 *Skull measuring has a long history.* See Bindman (2002) pp.201–21 for Camper on skull measurement (from which the quotes as well).

340 *Sadly, Camper's iconography.* See Gould (1981) for the classic debunking work on craniometry and IQ. See Lahr (1996), Hanihara (2002) and Hennessy and Stringer (2002) for recent major craniometric studies, all of which build on the work of Bill Howells.

341 *Human skulls are wonderfully diverse.* See Lahr (1996) for an authoritative treatment of recent human skull diversity. The relative prognathism of Eskimos and Australian Aborigines is calculated from Hanihara (2002) Table 3.

342 *My claim that we will soon be able.* Boas published several studies on his immigrant data set, the most important of which was Boas (1912). The Röse quote is from Boyd (1955) p.299. Two recent papers, Sparks and Jantz (2002) and Gravlee et al. (2003), have reanalysed Boas's data. The analysis done by each is somewhat different and they draw somewhat different conclusions. Sparks and Jantz (2002), however, do the critical analysis of variance – with ancestry, birthplace and their interaction as the effects. They show that there is a significant effect of birthplace and – just as one would expect from Boas's hypothesis – a strong interaction effect . Contrary to Boas, however, the plasticity is small compared to the persistence of ancestral effects and the interactions are not of the sort that would necessarily cause skull shape to converge. They do not accuse Boas of fraud, but one cannot help but suspect that he presented those results that favoured his hypothesis and ignored those that did not. Gould (1981) p.108 cites Boas with approval.

347 *They are only the latest casualties.* See notes to Chapter VI for the history of the negritos as well as the various essays in McEwan et al. (1997) for a history of the Selk'nam and their legend and their fate.

348 *'Beauty,' says the philosopher.* See Scarry (2000) p.4 for beauty and the impulse to reproduce. See Plato, *The symposium* (trans. W. Hamilton. 1951. Penguin Books, Harmondsworth, UK) p.87 for the same. See Darwin (1871, 1981) Vol. 2 p.92 for the Argus pheasant. See Bindman (2002) for a survey of eighteenth-century aesthetic theory with respect to race. See Darwin (1871, 1981) Vol.2 pp.342–54 on the particularity of beauty.

351 *The universality of beauty's standard.* See Thornhill and Gangestad (1999) for a survey of the recent literature on facial attractiveness. See Perrett et al. (1994) for a classical study on the perception of female beauty. What Brazilians say is recorded (with delight) by the late William Hamilton in *The narrow roads of gene land* (2002, Oxford University Press, Oxford) Vol. 2 p.677. Many of the ideas about the meaning of beauty expressed in this chapter can be traced to Hamilton's writings.

353 *The effects of poor childhood nutrition.* For the genetics of the face see Winter (1996). For spontaneous abortion as an adaptation to eliminate defective embryos see Forbes (1997).

354 *Mutation is a game of chance.* See Crow (2000) for the number of deleterious mutations and a model of truncation selection.

356 *Beauty, Stendhal says.* Stendhal, *De l'amour* (Folio, Paris) p.59.

BIBLIOGRAPHY

Abbott, M.E. and J. Kaufmann. 1916. Double monster of Janus type: cephalothoracopagus monosymmetros cyclops synotis. *Bulletin of the International Association of Medical Museums* 6: 95–101

Abramowicz, M.J. et al. 1992. Identification of a mutation in the coding sequence of the human thyroid peroxidase gene causing congenital goitre. *Journal of Clinical Investigation* 90: 1200–4

Abu-Abed, S. et al. 2001. The retinoic acid-metabolising enzyme CYP26A1, is essential for normal hindbrain patterning, vertebral identity and development of posterior structures. *Genes and Development* 15: 226–40

Addison, J. 1721. *The battel* [sic] *of the pygmies and cranes.* London

Adelmann, H.B. 1936a. The problem of Cyclopia I. *Quarterly Review of Biology* 11: 161–82

Adelmann, H.B. 1936b. The problem of Cyclopia II. *Quarterly Review of Biology* 11: 284–304

Afzelius, B. 1976. A human syndrome caused by immotile cilia. *Science* 193: 317–19

Alberch, P. 1986. Possible dogs. *Natural History* 95: 4–8

Alberts, B. et al. 1994. *The molecular biology of the cell.* Garland, N.Y.

Albin, R.L. 1988. The pleiotropic gene theory of senescence: supportive evidence from human genetic disease. *Ethology and Sociobiology* 9: 371–82

Aldrovandi, U. 1642. *Monstrorum historia.* Bononiae

Ames, B.N. et al. 1993. Oxidants, antioxidants, and the degenerative diseases of ageing. *Proceedings of the National Academy of Sciences, USA* 90: 7915–22

Anon. 1829a. *Bulletin des sciences médicales* 18: 169–72

Anon. 1829b. *La Clinique* 1: 200

Anon. 1829c. *La Clinique* 1: 254–5

Appel, T. 1987. *The Cuvier–Geoffroy debate: French biology in the decade before Darwin* Oxford University Press, Oxford, UK

Aristotle. 1984. *The complete works of Aristotle: the revised Oxford translation* J. Barnes (ed.) Princeton University Press, Princeton, N.J.

Aterman, K. 1999. From Horus the child to Hephaestus who limps: a romp through history. *American Journal of Medical Genetics* 83: 53–63

Bachiller, D. et al. 2000. The organiser factors chordin and noggin are required for mouse forebrain development. *Nature* 403: 658–61

Bacon, F. 1620 (2000). *The new organon.* L. Jardine and M. Silverthorne (eds) Cambridge Unversity Press, Cambridge, UK

Baker, J.R. 1974. *Race.* Oxford University Press, N.Y.

Balemans, W. et al. 2001. Increased bone density in sclerosteosis is due to the deficiency of a novel secreted protein (SOST). *Human Molecular Genetics* 10: 537–43

Baljet, B. 1990. The cyclopic monsters of the Vrolik collection. *Actes du 5ᵉ colloque de conservateurs des musées d'histoire des sciences médicales* 66–78.

Baljet, B. and M.L. Öjesjö. 1994. Teratology in art or the Dysmorphology–Hieronymus Bosch Connection. *Actes du 7ᵉ colloque des conservateurs des musées d'histore des sciences médicales*

Baljet, B. and R.–J. Oostra. 1998. Historical aspects of the study of malformations in the Netherlands. *American Journal of Medical Genetics* 77: 91–9

Baljet, B. et al. 1991. Willem Vrolik on cyclopia. *Documenta Opthalmologica* 77: 355–68

Bamshad, M. et al. 1999. Reconstructing the history of human limb development: lessons from birth defects. *Pediatric Research* 45: 291–9

Barbin, H. 1980. *Herculine Barbin: being the recently discovered memoirs of a nineteenth century French hermaphrodite* (intro. M. Foucault, trans. R. McDougall). Pantheon, N.Y.

Barbujani, G. et al. 1997. An apportionment of human DNA diversity. *Proceedings of the National Academy of Sciences USA* 94: 4516–19

Barinaga, M. 1993. New protein appears to be long-sought neural inducer. *Science* 262: 653–54

Barrett, D. and M. McCann. 1980. Two-toed man. *Sunday Times Magazine* 28–31

Barsh, G. 1999. Of ancient tales and hairless tails. *Nature Genetics* 22: 315–16

Bartke, A. et al. 2001a. Longevity – extending the lifespan of long-lived mice. *Nature* 414: 412

Bartke, A. et al. 2001b. Prolonged longevity of hypopituitary dwarf mice. *Experimental Gerontology* 36: 21–8

Bates, A.W. 2001. The *De monstrorum* of Fortunio Liceti: a landmark of descriptive teratology. *Journal of Medical Biography* 9: 49–54

Bateson, W. 1894. *Materials for the study of variation.* Macmillan, London

Bazopoulou-Kyrkanidou, E. 1997. What makes Hephaestus lame? *American Journal of Medical Genetics* 72: 144–55

Beckman, K.B. and B.N. Ames. 1998. The free radical theory of ageing matures. *Physiological Reviews* 78: 547–81

Beddington, R.S.P. 1994. Induction of a second neural axis by the mouse node. *Development* 120: 613–20

Beddington, R.S.P. and E.J. Robertson. 1999. Axis development and early asymmetry in mammals. *Cell* 96: 195–209

Bell, A.G. 1918. *The duration of life and conditions associated with longevity.* Genealogical Record Office, Washington, DC

Bendyshe, T. 1864. On the anthropology of Linnaeus. *Memoires of the Anthropological Society of London* 1: 421–58

Besser, G.M. and M.O. Thorner. 1994. *Clinical endocrinology* (2nd ed.) Ch. 11. Mosby-Wolfe, London

Betts, D.H. et al. 2001. Reprogramming of telomerase activity and rebuilding of telomere length in cloned cattle. *Proceedings of the National Academy of Sciences, USA* 98: 1077–82

Bindman, D. 2002. *Ape to Apollo: aesthetics and the idea of race in the eighteenth century.* Reaktion, London

Birkett, J. 1858. Congential, supernumerary and imperfectly developed auricles on the sides of the neck. *Transactions of the Pathological Society of London* 9: 448–9

Bittles, A.H. 2001. Consaguinity and its relevance to clinical genetics *Clinical Genetics* 60: 89–90

Blair, H.C. 1998. How the osteoclast degrades bone. *Bioessays* 20: 837–46

Blanchard, R. 1907. Nouvelles observations sur les nègres pies. Geoffroy Satin-Hilaire à Lisbonne. *Bulletin de la Société Française d'histoire de la médecine* 6: 111–35

Blasco, M.A. et al. 1997. Telomere shortening and tumor formation by mouse cells lacking telomerase RNA. *Cell* 91: 25–34

Boaistuau, P. 1560 (2000) *Histoires prodigieuses*, Facsimile of Wellcome MS 136. (S. Bamforth ed.) Franco Maria Ricci, Milan

Boardman, J. 1997a. Pan in text pp.923–41; plate pp.612–36 in *Lexicon iconographicum mythologiae classicae*: VIII: 1 (text); 2 (Plates)

Boardman, J. 1997b. *The great god Pan.* Thames and Hudson, London

Boas, F. 1912. Changes in bodily form of descendants of immigrants. *American Anthropologist* 14: 530–62

Bodnar, A.G. et al. 1998. Extension of life-span by introduction of telomerase into normal human cell. *Science* 279: 349–51

Bogin, B. 1999. *Patterns of human growth*, 2nd ed. Cambridge University Press, Cambridge. UK

Bonaventure, J. et al. 1996. Common mutations in the Fibroblast Growth Factor Receptor 3 (FGFR3) gene account for achondroplasia, hypochondroplasia and thanatophoric dwarfism. *American Journal of Medical Genetics* 63: 148–54

Bondeson, J. 1997. *A cabinet of medical curiosities*. Tauris, London

Bondeson, J. 2000. *The two-headed boy and other medical marvels*. Cornell University Press, Ithaca. N.Y.

Bondeson, J. and A.E.W. Miles. 1996. The hairy family of Burma: a four generation pedigree of congenital hypertrichosis lanuginosa. *Journal of the Royal Society of Medicine* 89: 403–8

Bornstein, P.E. and R.R. Peterson. 1966. Numerical variation in the pre-sacral vertebral column in three population groups. *American Journal of Physical Anthropology* 25: 139–46

Boruwlaski, J. 1792. *The Memoirs of the celebrated dwarf, Joseph Boruwlaski, A Polish gentleman containing a faithful and curious account of his birth, education, marriage, travels and voyages*. (trans. S. Freeman) 2nd ed. J. Thompson. Birmingham

Boulet and Capecchi. 1996. Targeted disruption of Hoxc-4 causes esophageal defects and vertebral transformation. *Developmental Biology* 177: 232–49

Boyd, W.C. 1955. *Genetics and the races of man*. Little, Brown and Co., Boston

Brandt, A. 1897. Ueber die sogenannten Hundemenschen, beziehungsweise über Hypertrichosis universalis. *Biologische Zentralblatt* 17: 161–79

Broberg, G. 1983. *Homo sapiens*. Linnaeus' classification of man. in T. Frängsmyr, (ed.) *Linnaeus: the man and his work*. University of California Press, Berkley

Brockes, J.P. 1998. Regeneration and cancer. *Biochimica et biophysica acta*. 1377 M1–M11

Brown-Borg, H.M. et al. 1996. Dwarf mice and ageing process. *Nature* 384: 33

Browne, T. 1904. *The works of Thomas Browne*. C. Sayle (ed.) Grant Richards, London

Brueckner, M. 2001. Cilia propel the embryo in the right direction. *American Journal of Medical Genetics* 101: 339–44

Brunet, L.J. et al. 1998. Noggin, cartilage morphogenesis, and joint formation in the mammalian skeleton. *Science* 280: 1455–7

Brunkow, M.E. et al. 2001. Bone dysplasia sclerosteosis results from loss of the SOST gene product, a novel cystine knot-containing protein. *American Journal of Human Genetics* 68: 577–89

Bruyn, G.W. and L.N. Went 1986. Huntington's Chorea. in Vinken, G. W. et al. (eds.) *Extrapyramidal disorders: handbook of clinical neurology* 49: 267–73

Buffon, G.L. 1777. *Histoire naturelle générale et particulière*. Imprimerie Royale, Paris

Burn, J. and T. Strachan. 1995. Human embryo use in developmental research. *Nature Genetics* 11: 3–6

Callahan, C.A. and A.E. Oro. 2001. Monstrous attempts at adnexogenesis:

regulating hair follicle progenitors through sonic hedgehog signalling. *Current Opinion in Genetics and Development* 11: 541–6

Cargill, M. et al. 1999. Characterisation of single-nucleotide polymorphisms in coding regions of human genes. *Nature Genetics* 22: 231–8

Carpenter, E.M. et al. 1993. Loss of Hox-A1 (Hox-1. 6) function results in the reorganisation of the murine hindbrain. *Development* 118: 1063–75

Casey, B. and B.P. Hackett. 2000. Left–right axis malformations in man and mouse. *Current Opinon in Genetics and Development* 10: 257–61

Cavalli-Sforza, L.L. 1986. *The African pygmies*. Academic Press, N.Y.

Cavelaars, A.E.J.M. et al. 2000. Persistent variations in average height between countries and between socio-economic groups: an overview of 10 European countries *Annals of Human Biology* 27: 407–21

Celli, J. et al. 1999. Heterozygous germline mutations in the p53 homolog p63 are the cause of EEC syndrome. *Cell* 99: 143–51

Chapman, T. and L. Partridge. 1996. Female fitness in *Drosophila melanogaster*: an interaction between the effect of nutrition and of encounter rates with males. *Proceedings of the Royal Society, London Series B Biological Sciences* 263: 755–9

Charlesworth, B. 1996. Evolution of senescence: Alzheimer's disease and evolution. *Current Biology* 6: 20–2

Chen, L. et al. 2001. A Ser(365)->Cys mutation of fibroblast growth factor receptor 3 in mouse downregulates Ihh/PTHrP signals and causes severe achondroplasia. *Human Molecular Genetics* 10: 457–65

Chen, Y. et al. 2000. Conservation of early odontogenetic signaling pathways in Aves. *Proceedings of the National Academy, USA* 97: 10044–9

Chiang, C. et al. 1996. Cyclopia and defective axial patterning in mice lacking sonic hedgehog gene function. *Nature* 383: 407–12

Chiang, C. et al. 2001. Manifestation of the limb prepattern: limb development in the absence of sonic hedgehog function. *Developmental Biology* 236: 421–35

Chisaka, O. and M.R. Capecchi. 1991. Regionally restricted developmental defects resulting from targeted disruption of the mouse homeobox gene hox-1. 5 (HoxA3). *Nature* 350: 473–9

Cibelli, J.B. et al. 2002. The health profile of cloned animals. *Nature Biotechnology* 20: 13–14

Clark, R.M. et al. 2001. Reciprocal mouse and human limb phenotypes caused by gain and loss-of-function mutations affecting Lmbr1. *Genetics* 159: 715–26

Coates, M.I. and J.A. Clack. 1990. Polydactyly in the earliest tetrapod limbs. *Nature* 347: 66–9

Cockayne, E.A. 1933. *Inherited abnormalities of the skin and its appendages*. Oxford University Press, London

Cohen, M.M. 2002. Bone morphogenetic proteins with some comments on fibrodysplasia ossificans progressiva. *American Journal of Medical Genetics* 109: 87–92

Cohen, M.M. 1988. Further diagnostic thoughts about the Elephant Man. *American Journal of Medical Genetics* 29: 777–82

Cohen, M.M. 1989. A comprehensive and critical assessment of overgrowth and overgrowth syndromes. *Advances in Human Genetics* 18: 181–303

Cohen, M.M. 1993. Proteus syndrome: clinical evidence for somatic mosaicism and selective review. *American Journal of Medical Genetics* 47: 645–52

Cohen, M.M. 2001. Problems in the definition of holoprosencephaly. *American Journal of Medical Genetics* 103: 183–7

Cohen, M.M. and K. Shiota. 2002. Teratogenesis of holoprosencephaly. *American Journal of Medical Genetics* 109: 1–15

Cohn, M.J. and C. Tickle. 1999. Developmental basis for limblessness and axial patterning in snakes. *Nature* 399: 474–9

Cohn, M.J. and P.E. Bright. 2000. Development of vertebrate limbs: insight into pattern, evolution and dysmorphogenesis. in O'Higgins, P. and M.J. Cohn (eds) *Development, growth and evolution: implications for the hominid skeleton*. Academic Press, N.Y.

Collins-Schramm, H.E. 2002. Ethnic-difference markers for use in mapping by admixture linkage disequilibrium. *American Journal of Human Genetics* 70: 737–50

Colvin, J.S. et al. 1996. Skeletal overgrowth and deafness in mice lacking fibroblast growth factor receptor 3. *Nature Genetics* 12: 391–7

Comfort, A. 1964. *Ageing: the biology of senesence*. Holt, Rinehart and Winston. N.Y.

Conner, M. and M. Ferguson-Smith. 1993. *Essentials of medical genetics*. (5th ed.) Blackwell Science, Oxford

Conte, F. A. et al. 1994. A syndrome of female pseudohermaphroditism, hypergonadotropic hypogonadism, and multicystic ovaries associated with missense mutations in the gene encoding aromatase (P450 arom). *Journal of Clinical Endocrinology and Metabolism* 78: 1287–92

Corcoran, J. 1998. What are the molecular mechanims of neural tube defects? *Bioessays* 20: 6–8

Corder, E.H. et al. 1993. Gene dose of apololipoprotein E Type 4 allele and the risk of Alzheimer's disease in late onset families. *Science* 261: 921–3

Cornaro, L. 1550 (1903). *The art of living long: a new and improved English version of the treatise by the celebrated Venetian centenarian, Louis Cornaro with Essays by Joseph Addison, Lord Bacon and Sir William Temple*. W.F. Butler, Milwaukee

Cotsarelis, G. and S.E. Millar. 2001. Towards a molecular understanding of hair loss and its treatment. *Trends in Molecular Medicine* 7: 293–301

Crackower, M.A. et al. 1998. Defect in the maintenance of the apical ectodermal ridge in the Dactylaplasia mouse. *Developmental Biology* 201: 78–89

Crawfurd, J. 1827. *Journal of an embassy from the Governor-General of India to the court of Ava in the year 1827*. V.1.H. Colburn. London

Crosby, J.L. et al. 1992. Disorganisation is a completely dominant gain-of-function mouse mutation causing sporadic developmental defects. *Mechanisms of Development* 37: 121–6

Crow, J.F. 2000. The origins, patterns and implications of human spontaneous mutation. *Nature Reviews Genetics* 1: 40–7

Danerow, H. 1830. Ueber Ritta-Christina und die Siamesen. *Litterarishcen Annaleen der gesammten Heilkunde* 16: 454–82

Darwin, C. 1859 (1968). *The origin of species by means of natural selection.* Penguin, Harmondsworth, UK

Darwin, C. 1871 (1981). *The descent of man, and selection in relation to sex.* Princeton University Press, Princeton, N.J.

Darwin, C. 1882. *The variation of animals and plants under domestication*, 2nd ed. John Murray, London

Dasen, V. 1993. *Dwarfs in ancient Egypt and Greece.* Clarendon Press, London

Dasen, V. 1994. Pygmaioi. text pp.594–601; plates pp.466–86. *Lexicon Iconographicum Mythologiae Classicae* VIII: 1 (text); 2 (plates)

Dasen, V. 1997. Multiple births in Graeco–Roman antiquity. *Oxford Journal of Archaeology* 16: 49–61

Dasen, V. 2002. Les jumeaux siamois dans l'Antiquité classique: du mythe au phénomène de foire. *La Revue du Practicien* 52: 9–12

Daston, L. and K. Park. 1998. *Wonders and the order of nature 1150–1750.* Zone, N.Y.

De Luca, F. and J. Baron. 1999. Control of bone growth by fibroblast growth factors. *Trends in Endocrinology and Metabolism* 10: 61–5

De Quatrefages, A. 1895. *The pygmies* (trans. F. Starr). Macmillan, London

De Vijlder, J. et al. 1999. Defects in thyroid hormone supply. Ch. 16b. *The Thyroid and Its Diseases*. http://www.thyroidmanager.org

Del Campo, M. et al. 1999. Mondactylous limbs and abnormal genitalia are associated with hemizygosity for the human 2q31 region that includes the HOXD cluster. *American Journal of Human Genetics* 65: 104–10

Delange, F. and B. Hetzel. 2000. The Iodine Deficiency Disorders. Ch. 20. *The Thyroid and Its Diseases*. http://www.thyroidmanager.org

DeRobertis, E. and Y. Sasai. 1996. A common plan for dorsoventral patterning in Bilateria. *Nature* 380: 37–40

Diamond, J.M. 1982. Big-bang reproduction and ageing in male marsupial mice. *Nature* 298: 115–16

Diamond, J.M. 1991. Why are pygmies so small? *Nature*. 354: 111–12

Didde, R. 2002. Wetenschap. *Volkskrant* 25 May 2002

Dikötter, F. 1992. *The discourse of race in modern China*. Hurst, London

Dikötter, F. 1997. Hairy barbarians, furry primates, and wild men: medical science and cultural representations of hair in China. in Hiltebeitel, A. and B. D. Miller (eds) *Hair: its power and meaning in Asian cultures.* SUNY Press, Albany, N.Y.

Dikötter, F. 1998. *Imperfect conceptions: medical knowledge, birth defects, and eugenics in China.* Hurst, London

Dobson, J. 1958. Marie Sabina, the variegated damsel. *Annals of the Royal College of Surgeons* 22: 273–8

Dollé, P. et al. 1993. Disruption of the Hoxd-13 gene induces localised heterochrony leading to mice with neotenic limbs. *Cell* 75: 431–41

Donnai D and R.M. Winter 1989. Disorganisation: a model for 'early amnion rupture'? *Journal of Medical Genetics* 26: 421–5

Dreger, A.D. 1998. *Hermaphrodites and the medical invention of sex.* Harvard University Press. Cambridge, Mass.

Dudley, A.T. et al. 2002. A re-examination of proximodistal patterning during vertebrate development. *Nature* 418: 539–44

Durham-Pierre, D. et al. 1994. African origin of an intragenic deletion of the human P gene in tyrosine positive oculocutaneous albinism. *Nature Genetics* 7: 176–9

Egenvall, A. et al. 2000. Age pattern of mortality in eight breeds of insured dogs in Sweden. *Preventative Veterinary Medicine* 46: 1–14

Eigenmann, J.E. 1987. Insulin-like growth factor 1 in dogs. *Frontiers of Hormone Research* 17: 161–72

Eigenmann, J.E. et al. 1988. Insulin-like growth factor 1 levels in proportionate dogs, chondrodystrophic dogs and in giant dogs. *Acta Endocrinologica (Copenhagen)* 118: 105–8

Eigenmann, J.E. et. al. 1984. Body size parallels insulin-like growth factor 1 levels but not growth hormone secretory capacity. *Acta Endocrinologica (Copenhagen)* 106: 448–53

Emery, A.E.H. 1996. Genetic disorders in portraits. *American Journal of Medical Genetics* 66: 334–9

Epstein, C.J. et al. 1987. Transgenic mice with increased Cu/Zn-superoxide dismutase activity: animal model of dosage effects in Down syndrome. *Proceedings of the National Academy of Sciences USA* 84: 8044–8

Eyre-Walker, A. and P.D. Keightley. 1999. High genomic deleterious mutation rates in hominids. *Nature* 397: 334–47

Farrell, H.B. 1984. The two-toed Wadoma – familial ectrodactyly in Zimbabwe. *South African Journal of Medicine* 65: 531–3

Fässeler, P.E. and K. Sander. 1996. Hilde Mangold (1898–1924) and Spemann's organiser: Achievement and Tragedy. *Wilhelm Roux Archives of Developmental Biology* 205: 323–32

Felgenhauer, W.–R. 1969. Hypertrichosis lagnuinosa universalis. *Journal de Génétique humaine* 17: 1–44

Ferretti, P. and C. Tickle. 1997. The limbs. in P. Thorogood (ed.) *Embryos, genes and birth defects*. John Wiley and Sons, N.Y.

Figuera, L.E. 1995. Mapping of the congenital generalised hypertrichosis locus to chromosome Xq24–q27. 1. *Nature Genetics* 10: 202–6

Finch, C.E. 1990. *Longevity, senescence and the genome*. Chicago University Press

Finch, C.E. and G. Ruvkun. 2001. The genetics of ageing. *Annual Reviews of Genomics and Human Genetics* 2: 435–62

Fischer, J.L. 1972. Le concept expérimental dans l'oeuvre tératologique d'Etienne Geoffroy Saint-Hilaire. *Revue d'histoire des sciences* 25: 347–62

Fischer, J.L. 1991. *Monstres: histoire du corps et de ses défauts*. Syros, Paris

Flanagan, N. et al. 2000. Pleiotropic effects of the melanocortin 1 receptor (MC1R) gene on human pigmentation. *Human Molecular Genetics*. 9: 2531–7

Flatt, A.E. 1994. *The care of congenital hand anomalies*. Quality Medical Publishing. St Louis. Mo.

Fleischer, R. 1984. Artemis Ephesia. text pp.755–63; plates pp.564–73 in *Lexicon iconographicum mythologiae classicae* II: 1 (text); 2 (plates)

Flurkey, K. et al. 2001. Lifespan extension and delayed immune and collagen aging in mutant mice with defects in growth hormone production. *Proceedings of the National Academy of Sciences USA* 98: 6736–41

Forbes, L.S. 1997. The evolutionary biology of spontaneous abortion in humans. *Trends in Ecology and Evolutionary Biology* 12: 446–50

Frank, L.G. 1997. Evolution of genital masculinisation: why do female hyenas have such a large 'penis'? *Trends in Ecology and Evolution* 12: 58–62

Frankcom, G. and J.H. Musgrave. 1976. *The Irish giant*. Duckworth, UK

Fraser, G. 1877. *Wigtown and Whithorn: historical and descriptive sketches, stories and anecdotes, illustrative of the racy wit and pawky humour of the district*. Gordon Fraser, Wigtown

Fraumeni, J.F. 1967. Stature and malignant tumors of bone in childhood and adolesence. *Cancer* 20: 967–73

Freire-Maia, A. 1975. Genetics of acheiropodia (the handless and footless families of Brazil). VI. Formal genetic analysis. *American Journal of Human Genetics* 27: 521–7

Freire-Maia, A. 1981. Historical note: The extraordinary handless and footless families of Brazil – 50 years of acheiropodia. *American Journal of Human Genetics* 9: 31–41

Frey, J. 1994. *Toulouse-Lautrec: a life*. Weidenfeld and Nicolson, London

Frey, J. 1995a. What dwarfed Toulouse-Lautrec? *Nature Genetics* 10: 128–30

Frey, J. 1995b. Toulouse-Lautrec's diagnosis – reply. *Nature Genetics* 11: 363

Friedman, J.B. 1981. *The monstrous races in medieval thought and art*. Harvard, Cambridge, Mass.

Fromental-Ramain, C. et al. 1996. Hoxa-13 and Hoxd-13 play a crucial role in the patterning of the autopod. *Development* 122: 2997–3011

Fullerton, S.M. et al. 2000. Apolipoprotein E variation at the sequence haplotype level: implications for the origin and maintenance of a major human polymorphism. *American Journal of Human Genetics* 67: 881–900

Galis, F. 1999. Why do almost all mammals have seven cervical vertebrae? Developmental constraints, Hox genes, and Cancer. *Journal of Experimental Zoology* 285: 19–26

Galis, F. 2001. Why five fingers? Evolutionary constraints on digit number. *Trends in Ecology and Evolution*. 16: 637–46

Garcia-Cruz D. et al. 2002. Inherited hypertrichoses. *Clinical Genetics* 61: 321–9

Gat, U. et al. 1998. De novo hair follicle morphogenesis and hair tumors in mice expressing a truncated Beta-catenin in skin. *Cell* 95: 605–14

Gates, R.R. 1961. The Melanesian dwarf tribe of Aiome, New Guinea. *Acta Geneticae Medicae et Gemellogiae*. 10: 277–311

Geffner, M.E. et al. 1995. Insulin-like growth factor 1 resistance in immortalised T cell lines from African Efe pygmies. *Journal of Clinical Endocrinology and Metabolism*. 80: 3732–8

Geffner, M.E. et al. 1996. IGF-1 does not mediate T lymphoblast colony formation in response to estradiol, testosterone, 1,25(OH)2, Vitamin D3 and triiodothyronine; studies in control and pygmy lines *Biochemical and Molecular Medicine* 59: 72–9

Gegenbauer, C. 1880. Critical remarks on polydactyly as atavism. *Morphologisches Jahrbuch* 6: 584–96

Gehring, W. and K. Ikeo. 1999. Pax 6: mastering eye morphogenesis and eye evolution. *Trends in Genetics* 15: 371–7

Geissler, W. et al. 1994. Male pseudohermaphroditism caused by mutations of testicular 17beta-hydroxysteroid dehydrogenase 3. *Nature Genetics* 7: 35–9

Gelb, D. et al. 1996. Pycnodysostosis, a lysosomal disease caused by Cathepsin K deficiency. *Science* 273: 1236–9

Gelfland, M., C.J. Roberts, and R.S. Roberts. 1974. A two-toed man from the Doma people of the Zambezi Valley. *Rhodesian History* 5: 92–5

Geoffroy Saint-Hilaire, E. 1822. *Philosophie anatomique des monstruosités humaines.* Deville-Cavellin, Paris

Geoffroy Saint-Hilaire, E. 1829. Rapport sur le monstre bicéphale Ritta-Christina. *Gazette de Santé* (No. 270)

Geoffroy Saint-Hilaire, E. 1838. *Notions synthétiques, historiques et physiologiques de Philosophie Naturelle.* Dénain, Paris

Geoffroy Saint-Hilaire, I. 1832–37. *Histoire générale et particulière des anomalies de l'organisation chez l'homme et les animaux.* J. B. Ballière, Paris

Geoffroy Saint-Hilaire, I. 1847. *Vie, travaux et doctrine scientifique d'Étienne Geoffroy Saint-Hilaire.* La Société Géologique de France, Paris

Gerhart, J. 2000. Inversion of the chordate body axis: Are their alternatives? *Proceedings of the National Academy of Sciences, USA* 97: 4445–8

Gertz, O. 1948. Artemis och hinden frontispisplanschen Linnes Fauna Svecica. *Svenska Linne-Sallskapets Arsskrift* 31: 13–37

Gilbert, S.F. 2000. *Developmental biology* (6th edition). Sinauer Associates, Sunderland, Mass.

Glass, B. 1947. Maupertuis and the beginning of genetics. *Quarterly Review of Biology* 22: 196–210

Glickman, S.E. et al. 1992. Hormonal correlates of 'masculinisation' in female spotted hyenas (*Crocuta crocuta*). 1. Infancy to sexual maturity. *Journal of Reproduction and Fertility* 95: 451–62

Goding, C.R. 2000. Mitf from neural crest to melanoma: signal transduction and transcription in the melanocyte lineage. *Genes and Development* 14: 1712–28

Gong, Y. et al. 1999. Heterozygous mutations in the gene encoding noggin affect human morphogenesis. *Nature Genetics* 21: 302–4

Goto, N. et al. 1987. The mexican hairless dog, its morphology and inheritance. *Experimental Animals (Tokyo)* 36: 87–90

Goujon, E. 1869. Étude d'un cas d'hermaphrodisme bisexuel imparfait chez l'homme. *Journal de l'anatomie et de la physiologie normales et pathologiques de l'homme et des animaux* 6: 599–616

Gould, G.M. and W. L. Pyle. 1897. *Anomalies and curiosities of medicine*. W.B. Saunders, Philadelphia

Gould, S.J. 1981. *The mismeasure of man*. W.W. Norton, N.Y.

Gould, S.J. 1983. *Hen's teeth and horse's toes*. W.W. Norton, N.Y.

Graves, J.A. 1998. Interactions between SRY and SOX genes in mammalian sex determination. *Bioessays* 20: 264–9

Gravlee, C.G. et al. 2003. Heredity, environment and cranial form: a reanalysis of Boas's Immigrant data. *American Anthropologist* 105: 125–38

Gruber, G.B. 1955. Historisches und aktuelles über das Sirenen-problem in der medizin. *Nova Acta Leopoldina* 17: 89–122

Gruman, G.J. 1966. A history of ideas about the prolongation of life. *Transactions of the American Philosophical Society* 56: 1–102

Gubbay, J. et al. 1990. A gene mapping to the sex-determining region of the mouse Y chromosome is a member of a novel family of embryonically expressed genes. *Nature* 346: 245–50

Guichard, C. et al. 2001. Axonemal dynein intermediate-chain gene (DNAI1) mutations result in situs inversus and primary ciliary dyskenesia (Kartagener syndrome). *American Journal of Human Genetics* 68: 1030–5

Günther, T. and T. Schinke. 2000. Mouse genetics have uncovered new paradigms in bone biology. *Trends in Metabolism*. 11: 189–93

Guyda, H.J. 1998. Growth hormone therapy for non-growth hormone-deficient children with short stature. *Current Opinion in Pediatrics* 10: 416–21

Haldane, J.B.S. 1941. *New paths in genetics*. George Allen & Unwin, London

Hale, F. 1933. Pigs born without eyeballs. *Journal of Heredity* 24. 105–6

Haliburton, R.G. 1891. *The dwarfs of Mount Atlas*. David Nutt, London

Haliburton, R.G. 1894. Survivals of dwarf races in the new world. *Proceedings of the American Association for the Advancement of Science* 14: 1–14

Hall, B.K. 1995. Atavisms and atavistic mutations. *Nature Genetics* 10: 126–7

Hamburger, V. 1947. Monsters in nature. *Ciba Symposia* 9: 666–83

Hamburger, V. 1988. *The heritage of experimental embryology: Hans Spemann and the organiser.* Oxford University Press

Hamilton, J.B. 1942. Male hormone stimulation is a prerequisite and an incitant in common baldness. *American Journal of Anatomy* 71: 451–80

Hanihara, T. 2000. Frontal and facial flatness of major human populations. *American Journal of Physical Anthropology* 111: 105–34

Haraguchi, R. et al. 2000. Molecular analysis of external genitalia formation: the role of fibroblast growth factor (FGF) genes during genital tubercle formation. *Development* 127: 2471–9

Harding, R.M. et al. 2000. Evidence for variable selective pressure at MC1R. *American Journal of Human Genetics* 66: 1351–61

Hardy, M.H. 1992. The secret life of the hair follicle. *Trends in Genetics* 8: 55–61

Harrison, R.G. 1924. Some unexpected results of the heteroplastic transplantation of limbs. *Proceedings of the National Academy of Sciences, USA* 10: 69–74

Harvey, R.G. and D.R. Brothwell. 1969. Biosocial aspects of Ainu hirsuteness. *Journal of Biological Sciences* 1: 109–24

Harvey, W. 1965. *The works of William Harvey*. Willis, R. (ed., trans.) Sydenham Society, London

Haupt, H. et al. 1990. *Le bestiaire de Rodolphe II.* (trans. L. Marcou) Citadelles, Paris

Hayashizaki, Y. et al. 1989. Thyroid-stimulating hormone (TSH) deficiency caused by a single base substitution in the CAGYC region of the β-subunit. *EMBO Journal* 8: 2291–6

Headon, D.J. and P.A. Overbeek. 1999. Involvement of a novel Tnf receptor homologue in hair follicle induction. *Nature Genetics* 22: 370–4

Healy, E. et al. 2001. Functional variation of MC1R alleles from red-haired individuals. *Human Molecular Genetics.* 10: 2397–2402

Heijmans, B.T. et al. 2000. Common gene variants, mortality and extreme longevity in humans. *Experimental Gerontology* 35: 865–77

Hennessy, R.J. and C.B. Stringer. 2002. Geometric morphometric study of the regional variation of modern human craniofacial form. *American Journal of Physical Anthropology* 117: 37–48

Herdt, G. 1994. Mistaken sex: culture, biology and the third sex in New Guinea. in G. Herdt (ed.) *Third sex, third gender.* Zone Books. Cambridge, Mass.

Heron, T.M. 1986. *Boruwlaski, the little count.* Durham, UK.

Hertel, C. 2001. Hairy issues: Portraits of Petrus Gonsalus and his family in Archduke Ferdinand II's *Kunstkammer* and their contexts. *Journal of the History of Collections* 13: 1–22

Hinchliffe, J.R. and D.R. Johnson. 1980. *The development of the vertebrate limb.* Claredon Press, Oxford

Holekamp, K.E. et al. 1996. Rank and reproduction in female spotted hyenas. *Journal of Reproduction and Fertility* 108 229–37

Holliday, R. 1989. Food, reproduction and longevity: is the extended lifespan of calorie-restricted animals an evolutionary adaptation? *Bioessays* 10: 125–7

Holly, J.M.P. et al. 1999. Growth hormone, IGF-1 and cancer. Less intervention to avoid cancer? More intervention to prevent cancer? *Journal of Endocrinology* 162: 321–30

Horan, G.S. et al. 1994. Homeotic transformation of cervical vertebrae in Hoxa-4 mutant mice. *Proceedings of the National Academy of Sciences, USA.* 91: 12644–8

Horan, G.S. et al. 1995a. Compound mutants for the paralogous Hoxa-4, Hoxb-4, and Hoxc-4 genes show more complete homeotic transformations and a dose dependent increase in the number of vertebrae transformed. *Genes and Development* 9: 1667–77

Horan, G.S. et al. 1995b. Mutations in paralogous Hox genes result in overlapping homeotic transformations of the axial skeleton – evidence for unique and redundant function. *Developmental Biology* 169: 359–72

Houssay, F. 1937. *De la nature, des causes, des différences des monstres.* Editions Hippocrates, Paris

Hu, D. and J.A. Helms. 1999. The role of sonic hedgehog in normal and abnormal craniofacial morphogenesis. *Development* 126: 4873–84

Huelsken, J. et al. 2001. Beta-catenin controls hair follicle morphogenesis and stem cell differentiation in the skin. *Cell* 105: 533–45

Hummel, K.P. 1958. The inheritance and expression of Disorganization, an unusual mutation in the mouse. *Journal of Experimental Zoology* 137: 389–423

Hummel, K.P. 1959. Developmental anomalies in mice resulting from action of the gene Disorganization, a semi-dominant lethal. *Pediatrics* 23: 212–21

Ianakiev, P. et al. 2000. Acheiropodia is caused by a genomic deletion in C7orf2, the human orthologue of the Lmbr1 gene. *American Journal of Human Genetics* 68: 38–45

Imperato-McGinley, J. et al. 1974. Steroid 5-alpha-reductase deficiency in man: an inherited form of male pseudohermaphroditism. *Science* 186: 1213–15

Imperato-McGinley, J. et al. 1991. A cluster of male pseudohermaphrodites with 5-Alpha-reductase deficiency in Papua New Guinea. *Clinical Endocrinology* 34: 293–8

Incardona, J.P. 1998. The teratogenic *Veratrum* alkaloid cyclopamine inhibits sonic hedgehog signal transduction. *Development* 125: 3553–62

International Sequencing Consortium. 2001. Initial sequencing and analysis of the human genome. *Nature* 409: 860–921

Irving, J. 1862. *The drowned women of Wigton, a romance of the Covenant.* Porteous and Hislop, Glasgow

Isaac, A. et al. 2000. FGF and genes encoding transcription factors in early limb specification. *Mechanisms of Development* 93: 41–8

Iwai, T. 1907. A statistical study on the polymastia of the Japanese. *Lancet* 2: 753–4

Jackson, I.J. 1997. Homologous pigmentation mutations in human, mouse and other model organisms. *Human Molecular Genetics* 6: 1613–24

Jackson, W.P.U. 1951. Osteo-dental dysplasia (Cleidocranial dysostosis) 'The Arnold Head'. *Acta Medica Scandinavica* 139: 293–5

Janin, J. 1829. (1998) *Une femme à deux têtes.* S. Pestel (ed.) La collection électronique de la Bibliotèque Municipale de Lisieux. http://ourworld. compuserve.com./homepages/bibhhlisieux/

Jarvik, G.P. et al. 1994. Non-mendelian transmission in a human developmental disorder: Split Hand/Split Foot. *American Journal of Human Genetics* 55: 710–13

Jeannotte, L. et al. (1993) Specification of axial identity in the mouse: role of the Hoxa5 (Hox1. 3) gene. *Genes and Development* 7: 2085–96

Jenkins, P. 1998. Cancer in acromegaly. *Trends in Endocrinology and Metabolism* 9: 360–6

Johanson, D. and B. Edgar. 1996. *From Lucy to language.* Orion, London

Joseph, R. and P. Godson. 1988. Peace at last for tragic Rita: white outcase in black skin. *Sunday Times* (Johannesburg). 28 August. p. 12

Jost, A. 1946–47. Recherches sur la différenciation sexuelle de l'embryon de lapin (Troisième Partie). *Archives d'anatomie microscopique et de morphologie expérimentale* 36: 271–315

Jung, H.–S. et al. 1998. Local inhibitory action of BMPs and their relationships with activators in feather formation: implications for periodic patterning. *Developmental Biology* 196: 11–23

Kappler, C. 1980. *Monstres, démons et merveilles à la fin du Moyen age.* Payot, Paris

Kaufman, M.H. and K.S. O'Shea. 1978. Induction of monozygotic twinning in the mouse. *Nature* 276: 707–8

Keith, A. 1911. An inquiry into the nature of the skeletal changes in acromegaly *Lancet* i: 993–1002

Kenyon, C. et al. 1993. A *C. elegans* mutant that lives twice as long as wild type. *Nature* 366: 461–4

Kere, J. et al. 1996. X-linked anhidrotic (hypohidrotic) ectodermal dysplasia is caused by a mutation in a novel transmembrane protein. *Nature Genetics* 13: 409–16

Kingdon-Ward, F. 1924. *From China to Hkamti Long*. Edward Arnold, London

Kingdon-Ward, F. 1937. *Plant hunter's paradise*. Jonathan Cape, London

Kirk, G.S. 1974. *The nature of the Greek myths*. Penguin, Harmondsworth, UK

Knussmann, R. et al. 1992. Relations between sex hormone level and characteristics of hair and skin in healthy young men. *American Journal of Physical Anthropology* 88: 59–67

Kobelt, G.L. 1844. The female sex organs in humans and some mammals (trans. H.F. Bernays) in Lowry, T.P. (1978) *The classic clitoris, historical contributions to scientific sexuality*. Nelson-Hall, Chicago

Kohn, M. 1995. *The race gallery: the return of racial science*. Jonathan Cape, London

Kollar, E.J. and C. Fisher. 1980. Tooth induction in chick epithelium: expression of quiescent genes for enamel synthesis. *Science* 207: 993–5

Komori et al. 1997. Targeted disruption of Cbfa1 results in a complete lack of bone formation owing to maturational arrest of osteoblasts. *Cell* 89: 755–64

Kondo, S. et al. 2001. The medaka rs-3 locus required for scale development encodes ectodysplasin-A receptor. *Current Biology* 7: 1201–6

Kondo, T. et al. 1997. Of fingers, toes, and penises. *Nature* 390: 29

Koren, Y. and E. Negev. 2003. *Im Herzen waren wir Riesen*. Econ, Munich

Kornak, U. et al. 2000. Mutations in the a3 subunit of the vacuolar H+-ATPase cause infantile malignant osteopetrosis. *Human Molecular Genetics* 9: 2059–63

Kostic, D. and M.R. Capecchi. 1994. Targeted disruptions of the murine HoxA-4 and HoxA-6 genes result in homeotic transformations of components of the vertebral column. *Mechanisms of Development* 46: 231–47

Kremer, H. et al. 1995. Male pseudohermaphroditism due to a homozygous missense mutation of the luteinising hormone receptor gene. *Nature Genetics* 9: 160–4

Krude, H. et al. 1998. Severe early onset obesity, adrenal insufficiency and red hair pigmentation caused by POMC mutations in humans. *Nature Genetics* 19: 155–7

Kruglyak, L. and D.A. Nickerson. 2001. Variation is the spice of life. *Nature Genetics* 27: 234–6

Krzisnik, C. et al. 1999. The 'Little People' of the Island of Krk – Revisited. Etiology of hypopituitarism revealed. *Journal of Endocrine Genetics* 1: 9–19

Kuester and Happle. 1984. The inheritance of common baldness. Two B or not two B? *Journal of the American Academy of Dermatology* 11: 921–6

Läärä, E. and P. Rantakallio. 1996. Body size and mortality in women: a 29-year follow up of 12,000 pregnant women in northern Finland. *Journal of Epidemiology and Community Health* 50: 408–14

Lahr, M.M. 1996. *The evolution of modern human diversity. a study in cranial variation.* Cambridge University Press

Lamb, T.M. et al. 1993. Neural induction by the secreted polypeptide noggin. *Science* 262: 713–18

Lammer, E.J. et al. 1985. Retinoic acid embryopathy. *New England Journal of Medicine* 313: 837–41

Landucci, L. 1542, 1927. *A Florentine diary from 1450 to 1516 by Luca Landucci, continued by an anonymous writer till 1542 with notes by Iodoco del Badia* (trans. A. de Rosen Jervis). J.M. Dent & Sons, London

Lanza, R.P. et al. 2000. Extension of cell life-span and telomere length in animals cloned from senescent somatic cells. *Science* 288: 665–8

Laqueur, T.W. 1989. 'Amor Veneris, vel Dulcedo Appeleteur' pp.90–131 in M. Feher (ed.) *Zone 5. Fragments for a history of the human body, part 3.* Zone, N.Y.

Laqueur, T.W. 1990. *Making sex, body and gender from the Greeks to Freud*, Harvard University Press, Cambridge. Mass.

Laue, L.L. et al. 1996. Compound heterozygous mutations of the luteinising hormone receptor gene in Leydig cell hypoplasia. *Molecular Endocrinology* 10: 987–97

Lawrence, P. 1992. *The making of a fly.* Blackwell. London

Laycock, J. and P. Wise. 1996. *Essential Endocrinology.* (3rd ed.) Oxford University Press

Lazner, F. et al. 1999. Osteopetrosis and osteoporosis: two sides of the same coin. *Human Molecular Genetics* 8: 1839–46

Le Guyader, H. 1998. *Étienne Geoffroy Saint-Hilaire (1772–1844): un naturalist visionnaire.* Belin, Paris

Le Mouellic, H. et al. 1992. Homeosis in the mouse induced by a null mutation in the Hox-3. 1 gene. *Cell* 69: 251–64

Le Roith, D. et al. 2001. What is the role of circulating IGF? *Trends in Endocrinology and Metabolism* 12: 48–52

Lee, C-K. et al. 1999. Gene expression profile of aging and its retardation by caloric restriction. *Science* 285: 1390–3

Lee, H.W. et al. 1998. Essential role of mouse telomerase in highly proliferative organs. *Nature* 392: 569–74

Lee, P.A. and S.F. Witchel. 1997. The influence of estrogen on growth. *Current opinion in pediatrics* 9: 431–6

Lenz, W. 1962. Thalidomide and congenital abnormalities. *Lancet* i: 45

Leroi, A.M. 2001. Molecular signals versus the *loi de balancement. Trends in Ecology and Evolution.* 16: 24–9

Leroi, A.M. et al. 2003. Cancer selection. *Nature Cancer Reviews* 3: 226–31

Lettice, L.A. et al. 2002. Disruption of a long-range cis-acting regulator for Shh causes preaxial polydactyly. *Proceedings of the National Academy of Sciences, USA* 99: 7548–53

Levin, M. et al. 1995. A molecular pathway determining left–right asymmetry in chick embryogenesis. *Cell* 82: 803–14

Levin, M. *et al.* 1996. Laterality defects in conjoined twins. *Nature* 384: 321

Lewandoski, M. et al. 2000. Fgf8 signalling from the AER is essential for normal limb development. *Nature Genetics* 26: 460–3

Lewis, E. 1978. A gene complex controlling segmentation in *Drosophila*. *Nature* 27: 565–70

Lewis, T. and D. Embleton. 1908. Split-hand and split-foot deformities, their types, origin, and transmission. *Biometrika* 6: 26–58

Li, J. et al. 1997. Dinucleotide repeat in the 3′ flanking region provides a clue to the molecular evolution of the Duffy gene. *Human Genetics* 99: 573–7

Liceti, F. 1634. *De monstrorum natura caussis et differentiis*. Padua

Licht, P. et al. 1992. Hormonal correlates of 'masculinisation' in female spotted hyenas (*Crocuta crocuta*). 2. Maternal and fetal steroids. *Journal of reproduction and fertility* 95: 463–74

Lickert, H. et al. 2002. Formation of multiple hearts in mice following deletion of beta-catenin in the embryonic endoderm. *Developmental Cell* 3: 171–81

Lifton, R.J. 1986. *The Nazi doctors: medical killing and the pychology of genocide*. Macmillan, London

Lindroth, S. 1983. The two faces of Linnaeus. in T. Frängsmyr (ed.) *Linnaeus: the man and his work*. University of California Press, Berkeley

Linnaeus, C. 1758, 1939. *Systema naturae* 10th edition. British Museum, London

Linnaeus, C. 1761. *Fauna svecica*. Stockholm

Lloyd, A.T. 1986. Pussy Cat, Pussy Cat, where have you been? *Natural History* 95(7): 46–52

Logroño, R. et al. 1997. Heteropagus conjoined twins due to fusion of two embryos: report and review. *American Journal of Medical Genetics* 73: 239–43

Lopez-Bermejo, A. et al. (2000). Genetic defects of the growth hormone-insulin-like growth factor axis. *Trends in Endocrinology and Metabolism* 11: 39–49

Lufkin, T. et al. 1991. Disruption of the Hox-1. 6 (Hoxa1) homeobox gene results in defects in a region corresponding to its rostral domain of expression. *Cell* 66: 1105–19

Lupu, F. et al. 2001. Roles of growth hormone and insulin-like growth factor 1 in mouse postnatal growth. *Developmental Biology* 229: 141–62

McBride, W.B. 1961. Thalidomide and congenital abnormalities *Lancet* ii: 1358

McEwan, C. et al. 1997. (eds) *Patagonia: natural history, prehistory and ethnography at the uttermost ends of the earth*. British Museum, London

Mackenbach, J.P. 1991. Narrowing inequalities in children's height. *Lancet* 338: 764

McLachlan, J. 1994. *Medical embryology*. Addison-Wesley, Wokingham

McLaren, A. 1990. What makes a man a man? *Nature* 346: 216–17

McMahon, J.A. et al. 1998. Noggin-mediated antatonism of BMP signalling is required for growth and patterning of the neural tube and somite. *Genes and Development* 12: 1438–52

McMullen, G. and K. Pearson. 1913. On the inheritance of the deformity known as split-foot or lobster-claw. *Biometrika* 9: 381–90

McPherron, A.C. and S.J. Lee. 1997. Doubling muscle in cattle due to mutations in the myostatin gene. *Proceedings of the National Academy of Sciences, USA* 94: 12457–61

McPherron, A.C. et al. 1997. Regulation of skeletal muscle mass in mice by a new TGF-Beta superfamily member. *Nature* 387: 83–90

Maden, M. 1999. Heads or tails? Retinoic acid will decide. *Bioessays* 21: 809–12

Maeder, T. 1998. A few hundred people turned to bone. *The Atlantic*, February. (two parts)

Manoiloff, E.O. 1931. A rare case of hereditary hexadactylism. *American Journal of Physical Anthropology* 15: 503–8

Manouvrier-Hanu S. et al. 1999. Genetics of limb anomalies in humans. *Trends in Genetics* 15: 409–17

Mansholt, U.J. 1987. The increase in the height of Dutchmen and the attraction of tennis. *Nederlands Tijdschrift voor Geneeskunde* 131: 376

Mark, M. et al. 1993. Two rhombomeres are altered in Hoxa-1 mutant mice. *Development* 119: 319–38

Maroteaux, P. 1995. Toulouse-Lautrec's diagnosis. *Nature Genetics* 11: 362

Maroteaux, P. and M. Lamy. 1965. The malady of Toulouse Lautrec. *JAMA, Journal of the American Medical Association* 191: 111–13

Marshall, H.K. and H.I. Harder. 1958. Testicular feminising syndrome in male pseudohermaphrodite: report of two cases in identical twins. *Obstetrics and Gynecology* 12: 284–93

Martin, E. 1880. *Histoire des monstres*. C. Reinwald, Paris

Martin, G. and J. Oshima. 2000. Lessons from human progeroid syndromes. *Nature* 408: 263–6

Martinez-Frias, M.-L. 1993. Another way to interpret the description of the Monster of Ravenna of the sixteenth century. *American Journal of Medical Genetics* 49: 362

Mascie-Taylor, C.G.N and J.L. Boldsen. 1985. Regional and social analysis of height variation in a contemporary British sample. *Annals of Human Biology* 12: 315–24

Mathon, N.F. et al. 2001. Lack of replicative sensecence in normal rodent glia. *Science* 291: 872–5

Mayor, A. 2000. *The first fossil hunters*. Princeton University Press, Princeton, N.J.

Medawar, P.B. 1952. *An unsolved problem in biology*. H. K. Lewis. London

Melanchthon, P. and M. Luther. 1523 (1823). *Interpretation of two horrible monsters [Deuttung der czwo grewlichï Figuren, etc.]*

Merimee, T.J. et al. 1981. Dwarfism in the pygmy. *New England Journal of Medicine* 305: 965–8

Merimee, T.J. et al. 1987. Insulin-like growth factors in pygmies: the role of puberty in determining final stature. *New England Journal of Medicine* 316: 906–11

Merke, F. 1993. *History and iconography of endemic goitre and cretinism*. MTP Press, Lancaster

Merlo, G.R. et al. 2002. Mouse model of split hand/foot malformation type 1. *Genesis* 33: 97–101

Meyers, E.N. and G.R. Martin. 1999. Differences in left–right axis pathways in mouse and chick: functions of FGF8 and SHH. *Science* 285: 403–6

Miller, R. and S. Austad. 1999. Large animals in the fast lane. *Science* 285: 199

Mills, A.A. et al. 1999. p63 is a p53 homologue required for limb and epidermal morphogenesis. *Nature* 398: 708–13

Mittwoch, U. 1973. *Genetics of sex differentiation*. Academic Press, N.Y.

Monreal, A.W. 1999. Mutations in the human homologue of mouse *dl* cause autosomal recessive and dominant hypohidrotic ectodermal dysplasia. *Nature Genetics* 22: 366–9

Montaigne, M. de. 1580 (1958). *Essays* (trans. J.M. Cohen). Penguin Books, Harmondsworth, UK

Montaigne, M. de. 1603 (1998). *Florio's translation of Montaigne's essays*. B. R. Schneider (ed.), Renascence Editions, University of Oregon

Moon, A.M. and M.R. Capecchi. 2000. Fgf8 is required for outgrowth and patterning of the limbs. *Nature Genetics* 26: 455–9

Morin, A. 1996. La teratologie de Geoffroy Saint-Hilaire à nos jours. *Bulletin de l'Association des Anatomistes* 80: 17–31

Morishima, A. et al. 1995. Aromatase deficiency in male and female siblings caused by a novel mutation and the physiological role of estrogens. *Journal of clinical endocrinology and metabolism* 80: 3689–98

Mortlock, D.P. and J.W. Innis. 1997. Mutation of Hox a-13 in hand-foot-genital syndrome. *Nature Genetics* 15: 179–80

Mortlock, D.P. et al. 1996. The molecular basis of hypodactyly (Hd): a deletion in Hox a-13 leads to arrest of digital arch formation. *Nature Genetics* 13: 284–8

Moskovitz, E. 1987. *By the grace of the devil*. Rotem, Ramat-Gan, Israel

Muenke M. and P.A. Beachy 2000. Genetics of ventral forebrain development and holoprosencephaly. *Current Opinion in Genetics and Development* 10: 262–9

Mundlos, S. 1999. Cleidocranial dysplasia: clinical and molecular genetics. *Journal of Medical Genetics* 36: 177–82

Mundlos, S. et al. 1997. Mutations involving the transcription factor CBFA1 cause cleidocranial dysplasia. *Cell* 89. 773–9

Muragaki, Y. et al. 1996. Altered growth and branching patters in synpolydactyly caused by mutations in Hox d-13. *Science* 272: 548–51

Mya-Tu, M. et al. 1962. Tarong pygmies in North Burma. *Nature* 195: 131–2

Mya-Tu, M. et al. 1966. *The Tarons in Burma*. Burma Medical Research Institute, Rangoon. Special Report Series No. 1

Nanni, L. et al. 1999. The mutational spectrum of the sonic hedgehog gene in holoprosencephaly: SHH mutations cause a significant proportion of autosomal dominant holoprosencephaly. *Human Molecular Genetics* 8: 2479–88

Nanni, L.et al. 2001. SHH mutation is associated with solitary median maxillary central incisor: a study of 13 patients and review of the literature. *Journal of Medical Genetics* 102: 1–10, 2001

Naski, M.C. et al. 1996. Graded activation of fibroblast growth factor receptor 3 by mutaitons causing achondroplasia and thanatophoric dysplasia. *Nature Genetics* 13: 233–7

Naski, M.C. et al. 1998. Repression of hedgehog signalling and BMP4 expression in growth plate cartilage by fibroblast growth factor receptor 3. *Development* 125: 4977–88

Naudin ten Cate L., C. Vermeij-Keers, D.A. Smit, T.W. Cohen-Overbeek, K.B. Gerssen-Schoorl, T. Dijkhuisen. 1995. Intracranial teratoma with multiple fetuses. Pre- and post-natal appearance. *Human Pathology* 26: 804–7

Neaves, W.B. et al. 1980. Sexual dimorphism of the phallus in spotted hyena (*Crocuta crocuta*). *Journal of Reproduction and Fertility* 59: 509–13

Needham, J. 1959. *A history of embryology*. Cambridge University Press, Cambridge, UK

Netter, A. et al. 1958. Le testicule feminisant. *Annales d' endocrinologie* 9: 994–1014

Neubert, R. et al. 1999. Developmental model for thalidomide action. *Nature* 400: 419–20

Newbery, H.J. and C.M. Abbott. 2002. Of mice, men and motor neurons. *Trends in Molecular Medicine* 8: 88–92

Niccoli, O. 1990. *People and prophecy in renaissance Italy*. (trans. L. G. Cochrane.) Princeton University Press, Princeton

Niswander, L. et al. 1993. FGF-4 replaces the apical ectodermal ridge and directs outgrowth and patterning of the limb. *Cell* 75: 579–87

Nonaka, S. et al. 1998. Randomisation of left–right asymmetry due to loss of nodal cilia generating leftward flow of extraembryonic fluid in mice lacking KIF3B motor protein. *Cell* 95: 839–47

Noramly, S. and B.A. Morgan. 1998. BMPs mediate lateral inhibition at successive stages in feather tract development. *Development* 125: 3775–87

O'Connell, H.E. et al. 1998. Anatomical relationship between urethra and clitoris. *Journal of Urology* 159: 1892–7

Olbrich, H. et al. 2002. Mutations in DNAH5 cause primary ciliary dyskinesia and randomisation of left–right asymmetry. *Nature Genetics* 30: 143–4

Olsen, B.R. et al. 2000. Bone development. *Annual Reviews of Cell and Developmental Biology* 16: 191–220

On-line Mendelian Inheritance in Man. 2000. *OMIM*™. McKusick-Nathans Institute for Genetic Medicine, Johns Hopkins University, Baltimore, MD, and National Center for Biotechnology Information, National Library of Medicine, Bethesda, MD. http://www. ncbi. nlm. nih. gov/omim/

Oosterhout, van C. et al. 2003. Inbreeding depression and genetic load of sexually selected traits: how the guppy lost its spots. *Journal of Evolutionary Biology* 16: 273–81

Oostra, R.-J. et al. 1998a. Congenital anomalies in the teratological collection of the Museum Vrolik in Amsterdam, the Netherlands. I: syndromes with multiple congenital anomalies. *American Journal of Medical Genetics* 77: 100–15

Oostra, R.-J. et al. 1998b. Congenital anomalies in the teratological collection of the Museum Vrolik in Amsterdam, the Netherlands. II: Skeletal Dysplasias. *American Journal of Medical Genetics* 77: 116–34

Oostra, R.-J. et al. 1998c. Congenital anomalies in the teratological collection of the Museum Vrolik in Amsterdam, the Netherlands. III: Primary field defects, sequences and other complex anomalies. *American Journal of Medical Genetics* 80: 46–59

Oostra, R.-J. et al. 1998d. Congenital anomalies in the teratological collection of the Museum Vrolik in Amsterdam, the Netherlands. IV: Closure defects of the neural tube. *American Journal of Medical Genetics* 80: 60–73

Oostra, R.-J. et al. 1998e. Congenital anomalies in the teratological collection of the Museum Vrolik in Amsterdam, the Netherlands. V: Conjoined and acardiac twins. *American Journal of Medical Genetics* 80: 74–89

Oro, A.E. and M.P. Scott. 1998. Splitting hairs: dissecting roles of signaling systems in epidermal development. *Cell* 95: 575–8

Orr, H.T. 2000. A proposed mechanism of ALS fails the test in vivo. *Nature Neuroscience* 5: 287–8

Ortega-Ortiz, J.G. and B. Villa-Ramirez. 2000. Polydactyly and other features of the manus of the vaquita, *Phocoena sinus*. *Marine Mammal Science* 16: 277–86

Paré, A. 1573 (1971). *Des monstres*. J. Céard (ed.) Droz, Geneva

Paré, A. 1573 (1982). *On monsters and marvels* (trans. J. L. Pallister) Chicago University Press, Chicago

Park, K. and L. Daston. 1981. Unnatural conceptions: the study of monsters in sixteenth and seventeenth century France and England. *Past and Present* 92: 20–54

Parkes, T.L. et al. 1998. Extension of *Drosophila* lifespan by overexpression of human SOD1 in motorneurons. *Nature Genetics* 19: 171–4

Partridge, L. and D. Gems. 2002. Mechanisms of ageing: public or private? *Nature Reviews Genetics* 3: 165–75

Patronek, G.J. et al. 1997. Comparative longevity of pet dogs and humans: implications for gerontology research. *Journal of Gerontology* 52A: B171–8

Pearson, K. et al. 1913. *A monograph on albinism in man.* 3 V. text; 3 V. plates. Draper's company research memoirs, Biometric series X. Dulau & Co. London

Pearson, K. 1908. On the inheritance of the deformity known as split-foot or lobster claw. *Biometrika* 6: 69–79

Pearson, K. 1913. Notes on the Honduras piebald. *Biometrika* 9: 330–1

Perrett, D.I. et al. 1994. Facial shape and judgements of female attractiveness. *Nature* 368: 239–42

Perriton, C. et al. 2002. Sonic hedgehog signalling from the urethral epithelium controls external genital development *Developmental Biology* 247: 26–46

Piccolo, S. et al. 1996. Dorsoventral patterning in *Xenopus*: inhibition of ventral signals by direct binding of Chordin to BMP-4. *Cell* 86: 589–98

Pinto-Correa, C. 1997. *The ovary of Eve: egg and sperm and preformationism.* Chicago University Press, Chicago

Pletcher, S.D. et al. 2002. Genome-wide transcript profiles in aging and calorically restricted Drosophila melanogaster. *Current Biology* 30: 712–23

Posel, D. 2001. Race as common sense: racial classification in twentieth century South Africa. *African Studies Review* 44: 87–113

Posner, G. L. and J. Ware. 1986. *Mengele: the complete story.* Futura, London

Power, C. and S. Matthews. 1997. Origins of health inequalities in a national population sample. *Lancet* 350: 1584–9

Qu, S. et al. 1998. Mutations in mouse Aristaless-like4 cause Strong's luxoid polydactyly. *Development.* 125: 2711–21

Quance, E. 1977. *Alexander Graham Bell, human inheritance, and the eugenics movement.* Research Bulletin of the National Historic Parks and Sites Branch, Parks Canada. No. 62

Quigley, C.A. et al. 1992. Complete deletion of the androgen receptor gene: definition of the null phenotype of the androgen insensitivity syndrome and determination of carrier status. *Journal of Clinical Endocrinology and Metabolism* 74: 932–3

Ramesar, R.S. et al. 1996.Mapping of the gene for cleidocranial dysplasia in the historical Cape Town (Arnold) kindred and evidence for locus homogeneity. *Journal of Medical Genetics* 33: 511–14

Ramirez-Solis et al. 1993. Hoxb-4 (Hox-2. 6) mutant mice show homeotic transformation of a cerivical rudiment and defects in the closure of the sternal rudiments. *Cell* 73: 279–94

Rana, B.K. et al. 1999. High polymorphism at the human melanocortin 1 receptor locus. *Genetics* 151: 1547–57

Rancourt et al. 1995. Genetic interaction between Hoxb-5 and Hoxb-6 is revealed by nonallelic noncomplementation. *Genes and Development* 9: 108–22

Reaume, A.G. et al. 1996. Motor neurons in Cu/Zn superoxide dismutase-deficient mice develop normally but exhibit enhanced cell death after axonal injury. *Nature Genetics* 13: 43–7

Reed, T.E. and J.V. Neel. 1959. Huntington's chorea in Michigan. *American Journal of Human Genetics* 11: 107–635

Reeves, R.H. et al. 2001. Too much of a good thing: mechanisms of gene action in Down syndrome. *Trends in Genetics* 17: 83–241

Reichert, H. and A. Simeone. 2001. Developmental genetic evidence for a monophyletic origin of the bilaterian brain. *Philosophical Transactions of the Royal Society B* 356: 1533–44

Reynolds, A.J. et al. 1999. Trans-gender induction of hair follicles. *Nature* 402: 46–7

Ricklefs, R.E. and C.E. Finch. 1995. Ageing: a natural history. *Scientific American*, N.Y.

Riddle, R.D et al. 1993. Sonic hedgehog mediates the Polarizing Activity of the ZPA. *Cell* 75: 1401–16

Robbins, L.S. et al. 1993. Pigmentation phenotypes of variant extension locus alleles result from point mutations that alter MSH receptor function. *Cell* 72: 827–34

Roberts, R.S. 1974. The making of a Rhodesian myth. *Rhodesian History* 5: 89–91

Robins, A.H. 1991. *Biological perspectives on human pigmentation*. Cambridge University Press, Cambridge, UK

Rodriguez, J.M. 1870. Descripcion de un monstruo cuadruple, nacido en Durango el ano de 1860. *Gaceta Medica de Mexico* 5: 33–48

Roessler, E. et al. 1996. Mutations in the human sonic hedgehog gene cause holoprosencephaly. *Nature Genetics* 14: 357–60

Root, A. 1998. Editorial: does growth hormone have a role in the management of children with nongrowth hormone deficient short stature and intrauterine growth retardation? *Journal of Clinical Endocrinology and Metabolism* 83: 1067–9

Rose, M.R. 1984. Laboratory evolution of postponed senescence in Drosophila melanogaster. *Evolution* 38: 1004–10

Rose, M.R. 1991. *Evolutionary biology of ageing*. Oxford University Press, N.Y.

Rosen, D.R. et al. 1993. Mutations in Cu/Zn superoxide dismutase gene are associated with familial amyotrophic lateral sclerosis. *Nature* 362: 59–62

Rosenbaum, S. et al. 1985. A survey of heights and weights of adults in Great Britain *Annals of Human Biology* 12: 115–27

Rosenbloom, A.L. and J.G. Guevara-Aguirre. 1998. Lessons from the genetics of Laron syndrome. *Trends in Endocrinology and Metabolism* 9: 276–83

Rosenburg, N.A. et al. 2002. Genetic structure of human populations. *Science* 298: 2381–5

Rosenfeld, R.G. et al. 1994. Growth hormone (GH) insensitivity due to primary GH deficiency. *Endocrine Reviews* 15: 369–90

Rousseau, F. et al. 1994. Mutations in the gene encoding fibroblast growth factor receptor-3 in achondroplasia *Nature* 371: 252–4

Rousseau, F. et al. 1995. Stop codon FGFR3 mutations in thanatophoric dwarfism type 1. *Nature Genetics* 10: 11–12

Rubinsztein, D.C. 2002. Lessons from animal models of Huntington's disease. *Trends in Genetics* 18: 202–9

Rudolph, K.L. et al. 1999. Longevity, stress response, and cancer in aging telomerase-deficient mice. *Cell*: 96: 701–12

Russell, A.J. et al. 1994. Mutation in the human gene for 3 alpha-hydroxysteroid dehydrogenase type II leading to male pseudohermaphroditism without salt loss. *Journal of Molecular Endocrinology* 12: 225–37

Sadler, T.W. 2000. *Langman's medical embryology* (8th ed.) Lippincott Williams & Wilkins, Philadelphia

Saegusa, H. et al. 1996. Targeted disruption of Hoxc-4 locus results in axial skeleton homeosis and malformation of the xiphoid process. *Developmental Biology* 174: 55–64

Saint-Ange, M. 1830. *Journal hebdomaidaire de médécine* 6: 42–9

Sakai, Y. et al. 20001. The retinoic acid-inactivating enzyme CYP26 is essential for establishing an uneven distribution of retinoic acid along the anterio-posterior axis within the mouse embryo. *Genes and Development* 15: 213–25

Samaras, T. and H. Elrick. 1999. Height, body size and longevity. *Acta Medica Okayama* 53: 149–69

Samaras, T. et al. 1999. Height, health and growth hormone. *Acta Paediatrica* 88: 602–9

Sandberg, D.E. et al. 1994. Short stature: a psychosocial burden requiring growth hormone therapy? *Pediatrics* 94: 832–9

Sato, N. et al. 1999. Induction of the hair growth phase in postnatal mice by localized transient expression of sonic hedgehog. *Journal of Clinical Investigation* 104: 855–64

Saunders, J.W. 1948. The proximo-distal sequence of origin of the parts of the chick wing and the role of the ectoderm. *Journal of Experimental Zoology* 108: 363–403

Saunders, J.W. and M.T. Gasseling. 1968. Ectodermal-mesenchymal interactions in the origin of limb symmetry. in R. Fleischmajer and R.F. Billingham (eds). *Epithelial-Mesenchymal interactions* pp.78–97. Williams and Wilkins, Baltimore

Scarry, E. 2000. *On beauty and being just*. Duckworth, London

Schächter, F. et al. 1994. Genetic associations with human longevity at the APOE and ACE loci. *Nature Genetics* 6: 29–32

Schatz, F. 1901. *Die Griechischen Götter und die Menschlichen Missgeburten*. J. F. Bergmann Verlag, Wiesbaden (reprint 1969, Editions Rodopi, Amsterdam)

Schebesta, P. 1952. *Die Negrito Asiens* 3 vols. Vienna

Schnaas, G. 1974. El Perro Pelon: mito, fantasia y biologia. *Gaceta medica de mexico* 108: 393–400

Schnitzer, E. 1888. *Emin Pasha in central Africa*. Schweinfurth, G., Ratzel, F. Felkin, R.W. Hartlauer, G. (eds) Philip and Son. London

Schweinfurth, G. 1878. *The heart of Africa* (trans. E. E. Frewer). 3rd edition. Sampson Low, Marston, Searle & Rivington, London

Segrave, K. 1996. *Baldness, a social history*. McFarland & Co. Jefferson, NC, USA

Sémonin, O. et al. 2001. Identification of three novel mutations of the noggin gene in patients with fibrodysplasia ossificans progressiva. *American Journal of Medical Genetics* 102: 314–17

Serres, E. 1832. *Recherches d'anatomie transcendante et pathologique. Théorie des formations et déformations organiques, appliquée à l'anatomie de Ritta-Christina, et de la duplicité monstrueuse*. J.B. Ballière, Paris

Seward, G.R. 1992. *The Elephant Man*. British Dental Association, London

Shriver, M.D. et al. 2003. Skin pigmentation, biogeographical ancestry and admixture mapping *Human Genetics* 112: 387–99

Sgrò, C. and L. Partridge. 1999. A delayed wave of death from reproduction in *Drosophila*. *Science* 286: 2521–4

Sharpe, P. 2001. Fish scale development: hair today, teeth and scales yesterday? *Current Biology* 11:R751–2

Sharpe, R.M. 1998. The roles of estrogen in the male. *Trends in Endocrinology and Metabolism* 9: 371–7

Shay, J.W. and W.E. Wright. 2000. Hayflick, his limit, and cellular ageing. *Nature Reviews Molecular Cell Biology* 1: 72–5

Shea, B.T and A.M. Gomez. 1988. Tooth scaling and evolutionary dwarfism: an investigation of allometry in human pygmies. *American Journal of Physical Anthropology* 77: 117–32

Shea, B.T. 1989. Heterochrony in human evolution: the case for neoteny reconsidered. *Yearbook of Physical Anthropology* 32: 69–101

Shea, B.T. and R.C. Bailey. 1986. Allometry and adaptation of body proportions and stature in African pygmies. *American Journal of Physical Anthropology* 100: 311–40

Shiels, P.G. et al. 1999. Analysis of telomere lengths in cloned sheep. *Nature* 399: 317

Shozu, M. et al. 1991. A new cause of female pseudohermaphroditism: placental aromatase deficiency. *Journal of Clinical Endocrinology and Metabolism* 72: 560–6

Shubin, N.H. and P. Alberch. 1986. A morphogenetic approach to the organization of the tetrapod limb. *Evolutionary Biology* 20: 319–87

Shubin, N.H. et al. 1997. Fossils, genes and the evolution of animal limbs. *Nature* 388: 639–48

Sidow, A. et al. 1999. A novel member of the F-box/WD40 gene family, encoding dactylin, is disrupted in the mouse dactylaplasia mutant. *Nature Genetics* 23: 104–7

Siebold, von C.T. 1878. Die Haarige familie von Ambras. *Archiv für Anthropolgie* 10: 253–60

Silventoinen, K. et al. 1999. Social background, adult body-height and health. *International Journal of Epidemiology* 28: 911–18

Siminoski, K. and J. Bain. 1993. The relationship among height, penile length and foot size. *Annals of Sex Research* 6: 231–5

Sinclair, A.H. et al. 1990. A gene from the human sex-determining region encodes a protein with homology to a conserved DNA-binding motif. *Nature* 346: 240–4

Slijper, E.J. 1942. Biologic-anatomical investigations on the bipedal gait and upright posture in mammals, with special reference to a little goat, born without forelegs. I., II. *Proceedings Koninklijke Nederlandse Academie van Wetenschap* 45: 288–95; 407–15

Smith, E.P. et al. 1994. Estrogen resistance caused by a mutation in the estrogen-receptor gene in a man. *New England Journal of Medicine* 331: 1056–61

Smith, R. et al. 1998. Melanocortin 1 receptor variants in an Irish population. *Journal of Investigative Dermatology* 111: 119–22

Sordino, P. et al. 1995. Hox gene expression in teleost fins and the origin of vertebrate digits. *Nature* 375: 678–81

Sornson, M.W. et al. 1996. Pituitary lineage determination by the Prophet of Pit-1 homeodomain factor defective in Ames dwarfism. *Nature*. 384: 327–32

Sparks, C.S. and R.L. Jantz. 2002. A reassessment of human cranial plasticity: Boas revisited. *Proceeding of the National Academy of Sciences USA* 99: 14636–9

Spemann, H. and H. Mangold 1924. Über die induktion von Embryonalanlagen durch Implantation artfremder Organisatoren. *Archiv für mikroskopische Anatomie und Entwicklungsmechanik* 100: 599–638

Spencer, R. 2000a. Theoretical and analytical embryology of conjoined twins: Part 1: Embryogenesis. *Clinical Anatomy* 13: 36–53

Spencer, R. 2000b. Theoretical and analytical embryology of conjoined twins: Part 2: Adjustments to union. *Clinical Anatomy* 13: 97–20

Spencer, R. 2001. Theoretical and analytical embryology of conjoined twins: Part 3: External, internal (fetuses in fetu and teratomas), and detached (Acardiacs). *Clinical Anatomy* 14: 428–44

Steinman, G. 2001a. Mechanisms of twinning. I. Effect of environmental diversity on genetic expression in monozygotic multifetal pregnancies. *The Journal of Reproductive Medicine* 46: 467–72

Steinman, G. 2001b. Laterality and intercellular bonding in monozygotic twinning. *The Journal of Reproductive Medicine.* 46: 473–9

Stephens, J.C. et al. 2001. Haplotype variation and linkage disequilibrium in 313 human genes. *Science* 293: 489–93

Stephens, T.D. et al. 2000. Mechanism of action in thalidomide teratogenesis. *Biochemical Pharmacology* 59: 1489–99

Stevens, G. et al. 1997. Oculocutaneous albinism (OCA2) in sub-Saharan Africa: distribution of the common 2.7-kb P gene deletion mutation. *Human Genetics* 99: 523–7

Stratakis, C.A. et al. 1998. The aromatase excess syndrome is associated with feminization of both sexes and autosomal dominant transmission of aberrant P450 aromatase gene transcription. *Journal of Clinical Endocrinology and Metabolism* 83: 1348–57

Sturm, R.A. et al. 1998. Human pigmentation genetics: the difference is only skin deep. *Trends in Genetics* 20: 712–21

Subramaniam, J.R. et al. 2002. Mutant SOD1 causes motor neuron disease independent of copper chaperone-mediated copper loading. *Nature Neuroscience* 5: 301–7

Sun, X. et al. 2002. Functions of FGF signalling from the apical ectodermal ridge in limb development. *Nature* 418: 501–8

Sutton, J.B. 1890. *Evolution and disease.* Walter Scott, London

Szabo, G. et al. 1969. Racial differences in the fate of the melanosomes. *Nature* 222: 1081–2

Tabin, C. 1998. A developmental model for thalidomide defects. *Nature* 396: 322–3

Tabin, C. 1999. Developmental model for thalidomide action – reply. *Nature* 400: 420

Takahashi, E. Secular trend in milk consumption and growth in Japan. *Human Biology* 56: 427–37

Ta-Mei, W. et al. 1982. Craniopagus parasiticus: a case report of a parasitic head protruding from the right side of the face. *British Journal of Plastic Surgery* 35: 304–11

Tang, D.G. et al. 2001. Lack of replicative senesence in cultured rat oligodendrocyte precusor cells. *Science* 291: 868–71

Tanner, J.M. 1981. *A history of the study of human growth*. Cambridge University Press, Cambridge, UK

Tanner, J.M. 1984. *Foetus into man*. (Revised ed.) Harvard University Press, Cambridge, Mass.

Tassabehji, M. et al. 1992. Waardenburg's syndrome patients have mutations in the human homologue of the Pax-3 paired box gene. *Nature* 355: 635–6

Tatsumi, K. et al. 1992. Cretinism with combined hormone deficiency caused by a mutation in the pit-1 gene. *Nature Genetics* 1: 56–8

Taussig, H.B. 1988. Evolutionary origin of cardiac malformations. *Journal of the American College of Cardiology* 12: 1079–86

Tavormina, P.L. et al. 1995. Thanatophoric dysplasia (types I and II) caused by distinct mutations in fibroblast growth factor receptor 3. *Nature Genetics* 9: 321–8

Thadini, K.I. 1934. The toothless men of Sind. *Journal of Heredity* 26: 65–6

Thangaraj, K. 2003. Genetic affinities of the Andaman Islanders, a vanishing human populations. *Current Biology* 13: 86–93

Thiery, M. and H. Houtzager. 1997. *Der Vrouwen Vrouwlijcheit*. Erasmus, Rotterdam

Thompson, A.A. and L.T. Nguyen. 2000. Amegakaryocytic thrombocytopenia and radio-ulnar synostosis are associated with HOXA11 mutation. *Nature Genetics* 26: 397–8

Thompson, C.J.S. 1930 (1994) *The history and lore of freaks*. Senate. London

Thornhill, R. and S.W. Gangestad 1999. Facial attractiveness. *Trends in Cognitive Science* 3: 452–60

Tibbles, J.A.R. and M.M. Cohen. 1986. The Proteus syndrome: the Elephant Man diagnosed. *British Medical Journal* 293: 683–5

Tickle, C. et al. 1975. Positional signalling and specification of digits in chick limb morphogenesis. *Nature* 20: 199–202

Tietze-Conrat, E. 1957. *Dwarfs and jesters in art*. Phaidon. London

Tjalma, R.A. 1966. Canine bone sarcoma: estimation of relative risk as function of body size. *Journal of the National Cancer Intitute*. 36: 1137–50

Toda, K. et al. 1972. Racial differences in melanosomes. *Nature New Biology* 236: 143–5

Ton, C.T. et al. 1991. Positional cloning and characterization of a Paired Box- and Homeobox-containing gene from the Aniridia region. *Cell* 67: 1059–74

Touchefeu-Meynier, O. 1992. Kyklops, Kyklopes. text pp.154–9; plates pp.69–75 in *Lexicon iconographicum mythologiae classicae*: VI: 1 (text); 2 (plates)

Townsend, P. et al.1992. *Inequalities in health*. Penguin Books, Harmondsworth, UK

Trotter, M. 1928. Hair growth and shaving. *Anatomical Record* 37: 373–9

Tsukui, T. et al. 1999. Multiple left–right asymmetry defects in Shh-/- mutant mice unveil a convergence of the Shh and retinoic acid pathways in the control of Lefty-1. *Proceedings of the National Academy of Sciences, USA.* 96: 11376–81

Twitty, V.C. 1966. *Of scientists and salamanders.* W. H. Freeman and Co., San Francisco

Twitty, V.C. and J.L. Schwind. 1931. The growth of eyes and limbs transplanted heteroplastically between two species of *Amblystoma* (sic) *Journal of Experimental Zoology* 59: 61–86

Tyson, E. 1699 (1966). *A philological essay concerning the pygmies of the ancients.* B. Windle (ed.). David Nutt, London

Unthan, C.H. 1935. *The armless fiddler: a pediscript.* George Allen & Unwin, London

Valenzano, M. et al. 1999. Sirenomelia. Pathological features, antenatal ultrasonographic clues, and a review of current embryogenic theories. *Human Reproductive Update* 5: 82–6

Valverde, P. et al. 1995. Variants of the melanocyte stimulating hormone receptor gene are associated with red hair and fair skin in humans. *Nature Genetics* 11: 328–30

Vassart, G. 2000. TSH receptor mutations and diseases Ch. 16a. *The Thyroid and Its Diseases.* http://www.thyroidmanager.org

Vieille-Grosjean, I. et al. 1997. Branchial Hox gene expression and human craniofaciall development. *Developmental Biology* 183: 4960

Viljoen, D.L. and P. Beighton. 1984. The split-hand and split-foot anomaly in a central African Negro population. *American Journal of Medical Genetics* 19: 545–52

Viljoen, D.L. and S.H. Kidson. 1990. Mirror-polydactlyly – pathogenesis based on a morphogen gradient theory. *American Journal of Medical Genetics* 35: 22935

Voss, J.W. and M.G. Rosenfeld. 1992. Anterior pituitary development. *Cell* 70: 527–30

Vrolik, W. 1834. Over den aard en oorsprong der cyclopie. *Niewe verhandelingen der Eerste Klasse van het Koninklijk Nederland Instituut.* 5: 25112

Vrolik, W. 1844–49. *Tabulae ad illustrandam embryogenesin hominis et mammalium tam naturalem quam abnormem.* Amsterdam, London

Waaler, H.T. 1984. Height, weight and mortality: the Norwegian experience. *Acta Medica Scandinavia Supplement* 679: 1–56

Walton, M.T. et al. 1993. Of monsters and prodigies: the interpretation of birth defects in the sixteenth century. *American Journal of Medical Genetics* 47: 7–13

Watanabe, A. et al. 1998. Epistatic relationship between Waardenburg syndrome genes MITF and PAX3. *Nature Genetics* 18: 283–6

Weber, G. 1995–99. *Lonely islands: the Andamanese.* The Andaman Association, Switzerland. http://andaman.org

Weinstein, B.S. and D. Ciszek. 2002. The reserve-capacity hypothesis: evolutionary origins and modern implications of the trade-off between tumor-suppression and tissue repair. *Experimental Gerontology* 37: 615–27

Westendorp, R.G.J. and T.B.L. Kirkwood. 1998. Human longevity and the cost of reproductive success. *Nature* 396: 743–6

Wilkie, A.O. et al. 1995. Apert syndrome results from localized mutations of FGFR2 and is allelic with Crouzon syndrome. *Nature Genetics* 9: 165–72

Williams, D. 1996. *Deformed discourse. The function of the monstrous in medieval thought.* University of Exeter Press

Williams, G.C. 1957. Pleiotropy, natural selection, and the evolution of senescence. *Evolution* 11: 398–411

Williams, G.R. 1998. Thyroid hormone action on cartilage and bone: interactions with other hormones at the epiphyseal plate and effects on linear growth. *Journal of Endocrinology* 157: 391–403

Williamson, S. and R. Nowak. 1998. The truth about women. *New Scientist.* 159: 34–5

Willier, B.H. and J.M. Oppenheimer. 1964. *Foundations of experimental embryology.* Prentice-Hall, N.J.

Wilmoth, J.R. et al. 2000a. Demography of longevity: past, present, and future trends. *Experimental Gerontology* 35: 1111–29

Wilmoth, J.R. et al. 2000b. Increase of maximum life-span in Sweden, 1861–1999. *Science* 289: 2366–86

Wilmut, I. 2002. Are there any normal cloned animals? *Nature Medicine* 8: 215–16

Wilmut, I. et al. 1997. Viable offspring derived from fetal and adult mammalian cells. *Nature* 385: 810–13

Wilson, D. 1993. *Signs and portents: monstrous births from the Middle Ages to the Enlightenment.* Routledge, London

Wilson, J.D. and C. Roehrborn. 1999. Long term consequences of castration in men: lessons from the Skoptzy and the Eunchs of the Chinese and Ottoman Courts. *Journal of Clinical Endocrinology and Metabolism* 84: 4324–31

Winter, R.M. 1996. What's in a face? *Nature Genetics* 12: 124-129

Winter, R.M. and D. Donnai. 1989. A possible human homologue for the mouse mutant disorganisation. *Journal of Medical Genetics* 26: 417–20

Wittkower, R. 1942. Marvels of the East. *Journal of the Warburg and Courtauld Institutes* 5: 159–97

Wolpert, L. 1971. Positional information and pattern formation. *Developmental Biology* 6: 183–224

Woolf, C.M. and F.C. Dukepoo. 1969. Hopi Indians, inbreeding, and albinism. *Science* 164: 30–7

Worden, G. 2002. *Mütter Museum.* Blast Books, N.Y.

Wright, S. 1935. A mutation of the guinea-pig, tending to restore the pentadactyl foot when heterozygous, producing a monstrosity when heterozygous. *Genetics* 20: 84–107

Wu, W. et al. 1998. Mutations in prop-1 cause familial combined pituitary hormone deficiency. *Nature Genetics* 18: 147–9

Yang, A. et al. 1999. p63 is essential for regenerative proliferation in limb, craniofacial and epithelial development. *Nature* 398: 714–18

Yang, Y. et al. 1997. Relationship between dose, distance and time in sonic hedgehog mediated regulation of anteroposterior polarity in the chick limb. *Development* 124: 4393–4404

Yu, C.E. et al. 1989. Positional cloning of the Werner's syndrome gene. *Science* 272: 258–62

Yule, H. 1858. *A narrative of the mission sent by the Governor-General of India to the court of Ava in 1855.* Bell, N.Y.

Zákány, J. et al. 1997. Regulation of number and size of digits by posterior Hox genes: A dose dependent mechanism with potential evolutionary implications. *Proceeding of the National Academy of Sciences, USA* 94: 1395–13700

Zapperi. R. 1995. Ein Haarmensch auf einem Gamälde von Agostino Carracci. in Hagner, M. (ed.) *Der falsche Körper: Bieträge zu einer Geschichte der Monstrositäten.* Wallstein, Göttingen

Zekraoui, L. et al. 1997. High frequency of the apolipoprotein ε4 allele in African pygmies and most of the African populations in sub-Saharan Africa. *Human Biology* 69: 575–81

Zeng, X. et al. 2001. A freely diffusible form of sonic hedgehog mediates long-range signalling. *Nature* 411: 716–20

Zguricas, J. et al. 1999. Clinical and genetics studies on 12 preaxial polydactlyly families and refinement of the localization of the gene responsible to a 1. 9 cM region on chromosome 7q35. *Journal of Medical Genetics* 36: 32–40

Zhou, X.–P. et al. 2000. Germline and germline mosaic PTEN mutations associated with a Proteus-like syndrome of hemihypertrophy, lower-limb asymmetry, arteriovenous malformations and lipomatosis. *Human Molecular Genetics* 19: 765–8

Zhou, X.–P. et al. 2001. Association of germline mutation in the PTEN tumor suppressor gene and Proteus and Proteus-like syndromes. *Lancet* 358: 210–11

Zimmerman, L.B. et al. 1996. The Spemann organizer signal noggin binds and inactivates Bone morphogenetic protein-4. *Cell* 86: 599–606

Zou, H. and L. Niswander. 1996. Requirement for BMP signalling in interdigital apoptosis and scale formation. *Science* 272: 738–41

INDEX

(Page references in *italic* refer to illustrations)

Gasseling, Mary 122–3
gastrula/gastrulation
36–7
Gegenbauer, Carl 132
Gehrig, Lou/Lou
Gehrig disease 316
Genara (pygmy queen)
181
genetic variety 336–40,
345; disappearance
of 346–7; in skulls
340–5, 342–3, 347
Geneviève (albino)
251, 252, 253, 254
genitalia 223–9;
ambiguous 235–6,
240; see pseudoher-
maphrodites; effect
of chromosomes on
230–3; malformed
14, 128; of spotted
hyenas 241–2
genome, human
15–16, 41, 42, 95,
101, 116
Geoffroy Saint-
Hilaire, Étienne
46–9, 62, 99–102
Geoffroy Saint-
Hilaire, Isidore 26,
47, 48, 169, 170;
Geranomachia (war of
the pygmies) 182, 184
Gérard, Marie/
Germain 236
gigantism, pituitary
178, 178–9, 207, 265
'glass bone disease' see
osteogenesis imper-
fecta
goats: with extra auri-
cles 84, 86; without
forelimbs 112

Gobineau, Arthur,
Comte de 262
Goethe, Johann von 49
goitres 196, 197, 198,
352
Goldschmidt, Richard
147
gonadotropin 303
gonads 18, 234; see also
testes
Gonsalvus, Arrigo
269–71, 270
Gonsalvus, Petrus 271,
272, 272–3, 284, 285
Gonsalvus, Tognina
271–2
Gould, Stephen Jay
344, 346
Goya, Francisco de
119
Gratton, George
Alexander 258, 259
Greek mythology 28,
68–9, 109, 181–3,
242; see also Artemis
Ephesia, Homer
Greek vases 167, 184–5
'growth hormone' 146,
175–6, 189, 203–4,
207, 212
growth rates, human
188–9, 202; and can-
cers 203–7; see also
height
Guanches, the 271
Guéneau de
Montbeillard, Count
Philibert 189
guevedoche see Salinas
Guilbert, Yvette 161,
164
guinea pigs: digits 121,
122

Gumilla, José: Orinoco
illustrado 258

H-Y antigens 232
hair 276–80, 281–2,
286, 287, 288; loss of
280–4; see also hairi-
ness, red-haired
people
hairiness/hirsutism/
hypertrichosis
lanuginosa 268–76,
270, 272, 277,
284–5, 286
Haldane, J.B.S. 298–9,
300, 301, 302
Hamburger, Viktor
147
Hamilton, James
282–3
Hands: ectrodactylous
107–9, 108, 111, 115;
missing see
acheiropody; see also
digits
Harvey, William 9–10;
De generatione ani-
malium 10, 51
heart attacks 16, 300,
329
hearts, extra 95–6
'hedgehog' genes 41,
74–5; 'sonic' 74,
75–7, 76, 124–5, 284
height 188–9, 211–13;
of eunuchs 201–2,
203; and longevity
208–9; and poverty
209–11; and testos-
terone receptors 235;
see also dwarfism,
gigantism, 'growth
hormone', pygmies

P.S.

Ideas,
interviews
& features...

The Shell Collector

Louise Tucker talks to Armand Marie Leroi

Where did you grow up?

I was born in Wellington, New Zealand, in 1964. I lived there until the age of five, then I moved to Pretoria, South Africa, where I lived until the age of fourteen, upon which I moved to Vancouver, Canada, where I lived until the age of … it becomes blurry now! … twenty-one.

What did your parents do that led you to travel so much?

My father was a Dutch diplomat. I am in fact a Dutch citizen and carry a Dutch passport. But I've never lived in Holland.

Would you want to?

Yes! When I make enough money I shall buy one of those tall, narrow, seventeenth-century canal-houses in Amsterdam. I think the Heerengracht would suit.

Which one of those countries is your favourite?

I loved South Africa very much; leaving was a wrench. But I've also become very attached to England.

What was your first memory of science? Was that at school?

I was very much a boy naturalist. I started when I was eleven years old or so, you know, watching birds, collecting rocks and shells and things. We've always been collectors in our family, natural – pathological – accumulators, and there was a bag of shells

that my parents had carted over from New
Zealand for whatever reason, shells that
they'd picked up on the beach.

**You carted them from New Zealand to
Pretoria?**
Yes, you see that's the way it is in diplomatic
families. Somebody else pays to move the
stuff so you never get rid of anything, you just
say 'Pack'. And then they pack and every time
you move to another country, another
container gets added onto all the others. So
we had this bag of shells – it was in a crate of
driftwood – and, well, I opened it up and I
began to classify them and before you know it
I had a shell collection. That's when I first
learnt Linnaeus' binomial system. By the time
I went to university I believed that my destiny
was to be *the* great authority on the land snails
of Africa.

The land snails.
Yes, specifically, the giant land snails of Africa.
Ultimately, however, my destiny was to write a
many-many-volumed monographic survey
of the entire terrestrial Mollusca of Africa.
Anybody who ever wanted to work on
terrestrial molluscs for the next hundred
years would have to refer to Leroi.

**How old were you when you were planning
that?**
I was seventeen or so. But then I went to
university. And, although malacology was a
good thing, it was clearly not cutting edge ▶

❝By the time I
went to university
I believed that my
destiny was to be
the great
authority on the
land snails of
Africa. ❞

3

Author photo © Jerry Bauer

LIFE
at a Glance

BORN

16 July 1964, Wellington,
New Zealand

NATIONALITY

Dutch

EDUCATED

Dalhousie University,
Canada (BSc Biology,
1989); University of
California, Irvine (PhD
Ecology and Evolutionary
Biology, 1993)

CAREER

Albert Einstein College of
Medicine, New York:
Postdoctoral Fellow 1993–6;
Division of Biology,
Imperial College London:
Lecturer 1996–2002;
Reader in Evolutionary
Developmental Biology
2002–present

AWARDS AND HONOURS

Times-Novartis Scientists
for the New Century
Medal, 2001; shortlisted,
Aventis prize for science
books, 2004; *Guardian*
First Book prize, 2004

The Shell Collector *(continued)*

◄ science. I became more interested in abstract questions of evolution processes. By the time I graduated and was considering my PhD, the shells had been relegated to their cabinet. I still have them though.

What is your current research?
I work in a field called 'evolutionary developmental biology' or 'evo-devo'. Scientists in this field want to understand the molecular basis of evolutionary change. We want to know the genes that are responsible, for example, for limb evolution: that enable a fish's fin to transform into a mouse's paw or a chicken's wing. Having said that, I work on worms. And they don't have any limbs at all.

Worms?
Yes. I work on this little worm, *C. elegans*. It's a very, very important worm – three guys got the Nobel Prize for their work on it a few years ago. I usually add that because people think it's strange that I work on a worm and I like to stress that it's quite an important worm, central to biological research. So, yes, that's what I still do. I work mainly on worms.

Where did the idea for this book come from?
Well, the way that developmental biologists study how an animal is put together is to look at mutants. It's a very routine thing. I look at mutant giant worms and mutant dwarf worms. I've got a laboratory full of them. Of course, the normal worm is only 1.2 millimetres long, and a giant worm is about 2 millimetres long, so it's all relative. Anyway, in the late 1990s, human genetics takes off. For a

long time paediatricians and human geneticists have spent their time classifying inherited diseases. Now, with the sequencing of the human genome, they can actually identify the mutant genes that cause these diseases. For the first time, we can use mutants to reverse-engineer humans, just as we use mutants to reverse-engineer worms. We can tell stories about the making of the human body. It was then that I realised that those stories could be put together in a book.

How long did it take?
An immensely long time. Of course I had a day job too. I got the contract at the beginning of 1999 so it took about five years.

And when did you write?
In the evenings and over the summer. One summer I went to Greece; I holed up in Lesvos. It was a good idea but I didn't actually get much writing done. Anyway, time ticks by; editors start getting mean; deadlines come and go; a final, final – perhaps there was even another final – deadline looms and the thing is done.

Did you find it difficult, that whole compartmentalising of time in order to find time to write?
Yes, I spent quite a bit of that time finding a voice, learning how to write. Now I think I could write it much faster, better. I hope so anyway.

And how did it feel to finish it?
Oh, it was very gratifying. In retrospect the delay, the time that it took, wasn't bad, it ▶

‘I work in a field called 'evolutionary developmental biology' or 'evo-devo'. Scientists in this field want to understand the molecular basis of evolutionary change. ’

The Shell Collector *(continued)*

◀ improved it. I fiddle endlessly with my text, I write and rewrite, and to some degree that's reflected in the book. It works a little bit better as a work of literature.

Do you think you developed your tone from learning how to communicate ideas to students or was it through the process of writing over a long time? How did you make it so literary?

It's certainly true that teaching imposes a discipline in terms of communicating material in the simplest way possible, with some force and clarity. But it must be said that the demands of writing are somewhat different from teaching. Writing a book demands a certain degree of narrative tension which is not required in teaching, and of course what makes writing fun, attractive, is that one can include cultural references beyond the immediate subject. Modern biological teaching doesn't accommodate that; you have to cut to the chase: there's a lot of material that has to be covered, and not much room for little cultural asides about the history and iconography of cyclopia. Students also get impatient, they want the facts.

In the process of creating this narrative, were you influenced by anyone? Were there any writers that you admired?

I have favourite stylists. I'm quite an enthusiast for W.G. Sebald and Bruce Chatwin. Among popular science writers I very much like Steve Jones. I think his *The Language of the Genes* is a masterwork. I read that book to bits: it's witty, it's clean,

6 I look at mutant giant worms and mutant dwarf worms. I've got a laboratory full of them. 9

6

information-dense and, in terms of the sheer clarity of ideas and results, hard to beat. I very much modelled my prose on his. Inevitably mine is different – not as good, but it's what I aimed at.

Had you ever written anything before?
I'd written a little bit for the *London Review of Books* and obviously lots of technical articles.

How did you feel about winning the *Guardian* First Book Award?
It was very flattering.

What's your guilty reading pleasure?
Patrick O'Brian's novels. I find Steven Maturin deeply appealing – he's a naturalist, an extremely good comparative anatomist who is not wanting in romantic and aesthetic sensibility either. A saturnine figure.

How do you write?
I carry a notebook wherever I go. I'm quite a fiddler so I print things out, stare at them with a pen in my hand, add a sentence, print it out again. I spend quite a lot of time staring at printout. ■

Top Ten
Favourite Reads

Gerald Durrell, *My Family and Other Animals*, 1956

Aristotle, *Historia Animalium*, trans. D'Arcy Wentworth Thompson, 1910

Vladimir Nabokov, *Lolita*, 1955

Patrick O'Brian, the Aubrey–Maturin series, 1970–99

C.P. Cavafy, *Collected Poems*, trans. John Mavrogordato, 1971

Amin Maalouf, *The First Century After Beatrice*, 1992

Sappho, trans. Mary Barnard, 1958

Frederick Rolfe, *The Desire and Pursuit of the Whole*, 1953

Karel Čapek, *War with the Newts*, 1936

Graham Greene, *The Quiet American*, 1955

Reading for *Mutants*

by Armand Marie Leroi

I BEGAN TO WRITE *Mutants* in January 1999. The book appeared (in the USA) in November 2003. From this you can infer, for it does not contain many words, that I did not find its writing easy. I had previously written many technical articles about genetics. That wasn't much help. The scientific literature has its own austere beauty, but it is also one in which cliché is an actual virtue. When I sat down to write *Mutants*, I had an idea, a contract and an advance. But I did not know what sort of book I wanted to write, or how to write it.

I did know what sort of book I did *not* want to write. Since the 1980s many popular science books have appeared that are written not by scientists but by journalists. Many are excellent. Others are full of this sort of thing:

> It was a blustery day when I walked up the imposing steps of the Rockefeller Institute. I was in search of the world's greatest expert on [fill in a horrific disease]. Dr Paul Weinstein was surprisingly young. His T-shirt said "Born to Clone", but he drove a Porsche. Fixing me with his clear blue eyes he said – and his voice broke with fatigue – "If we can find the gene, we can cure these kids."

It's a mode that has even crept into books written by scientists. Of course, scientists rarely make heroes of their colleagues. Instead, they do so of themselves. I blame the late Stephen Jay Gould for starting the rot. For my part, I was determined to do things differently. In my book there would be no homages to high-school biology teachers; no vignettes of exotic locales visited; no account of my own scientific triumphs (rightly so, they are modest enough); no personal anecdotes of my own struggle with deformity (I am bald, but I'm OK with that. Really).

My determination to have a remote, even austere authorial voice was not, incidentally, a decision that went down very well with agents and editors. Readers, they said, want to know who the author *is*; tell them, they urged, about *yourself*; you need a strong 'authorial presence'. Maybe so. But my life is, and has been, a quite conventional one. Readers, I reasoned, would need an unusual appetite for the mundane to find any of it interesting. If they want bathos mixed with their science, let them read Bill Bryson.

Some authors, when they write, do not read. Or at least they do not read in the genre that they are writing. They fear that their work will be contaminated by another's voice. Not me. Faced with the problem of how to write, I did the obvious thing: I went – as I always do – to my library.

I sought books that I would have liked to have written, took them down, and placed them on a separate shelf, heedless of subject. This is a partial list of those books:

Elaine Scarry, *On Beauty and Being Just*, 2002

James Davidson, *Courtesans and Fishcakes*, 1998

W.G. Sebald, *The Rings of Saturn*, 1998 ▶

❛When I sat down to write *Mutants*, I had an idea, a contract and an advance. But I did not know what sort of book I wanted to write, or how to write it. ❜

◀ Steve Jones, *The Language of the Genes*, 1993
Noel Annan, *Our Age*, 1990
Bruce Chatwin, *What Am I Doing Here*, 1989
Peter Levi, *The Hill of Kronos*, 1980
Cyril Connolly, *Enemies of Promise*, 1945
Walter Pater, *The Renaissance*, 1872
Stendhal, *On Love*, 1821

When stuck, confused or simply bored with my own writing, I would turn to these books, browse them and search for inspiration. I must have read certain passages dozens of times. Here is what I sought in them. In some I sought the solution to technical problems: in Annan, how to briskly review the tumultuous history of an idea; Sebald, how to write a long sentence; Scarry, how to write, indeed think, about beauty; Davidson, how to end a book. From others I sought more abstract qualities. Chatwin and Pater, compression and elision; Jones and Connolly, wit; Levi, humanity and scholarship; from Stendhal I sought candour. All of these writers are masters of expository prose. I have tried to emulate them, even be them, but I have always failed. ■

❝My determination to have a remote, even austere authorial voice was not, incidentally, a decision that went down very well with agents and editors. ❞

If You Loved This,
You Might Like…

Read on

Charles Darwin, *The Origin of Species*
Start at the beginning …

Paul Strathern, *Crick, Watson and DNA: The Big Idea*
Published in the Big Idea series, which aims to present complex material in a clear and accessible style, this is the history of Frances Crick's and James Watson's discovery of the double helix, the structure of DNA.

Steve Jones, *The Language of the Genes*
Bestselling account of how genetics can be used to explore human evolution.

Midas Dekkers, *The Way of All Flesh: A Celebration of Decay*
A counterblast to our current obsession with youth and defying age, Dekkers argues that to deny growing old is to deny life itself.

Matt Ridley, *Genome: The Autobiography of a Species in 23 Chapters*
The history of humanity, told through the story of a newly discovered gene in each of our 23 chromosomes.

Richard Dawkins, *Climbing Mount Improbable*
A journey through evolution, this book precedes the more famous *Selfish Gene* and starts the discussion of how our genes are in control of our destiny.

Richard Dawkins, *The Selfish Gene*
Dawkins rethinks the notion that we 'use' ▶

If You Loved This... *(continued)*

◀ our genes and argues that they 'use' us to reproduce.

Alessandro Boffa, *You're an Animal, Viskovitz!*
Funny anthropomorphic short tales in which the narrator, Viskovitz, tells his life and love story except in each tale he's a different creature. Learn about the love lives of snails, Chinese pigs and praying mantises.

Antony Doerr, *The Shell Collector*
Beautifully written and much-fêted début short story collection from the author of *About Grace*.

Jan Bondeson, *The Pig-faced Lady of Manchester Square: & Other London Medical Marvels*
An overview of seventeenth- and eighteenth-century attitudes to mutations such as dwarfism and giantism.

Sir Frederick Treves, *The Elephant Man and Other Reminiscences*
Treves is best known as the doctor who treated John Merrick (known as the Elephant Man: see WATCH ..., who suffered from a congenital disorder known as Proteus Syndrome), but was also the royal surgeon. See more about Treves at www.whonamedit.com/doctor.cfm/475.html and about the diagnosis at http://rarediseases.about.com/cs/proteussyndrome/a/031301.htm

Find Out More

VISIT

The Wellcome Library for the History and Understanding of Medicine
Has an immense collection of books and online resources.
210 Euston Road, London, NW1 2BE, UK
http://library.wellcome.ac.uk

Mütter Museum, Philadelphia
Part of the College of Physicians of Philadelphia, this museum has a large collection of anatomical and pathological specimens and medical illustrations.
19 South 22nd Street, Philadelphia, PA 19103, USA
www.collphyphil.org/muttpg1.shtml

SURF

www.armandleroi.com
The author's own website has information about his research and upcoming talks and appearances.

WATCH

Human Mutants
This three-part documentary series based upon *Mutants* was filmed in London, Amsterdam, Paris, Naples, Vienna, Salzburg, Cape Town and across the United States. Commissioned by Channel 4 (UK), produced by Tiger Aspect (London) and directed by award-winning British directors Philip Smith and Damon Thomas, it has also been shown in the USA on Discovery. Series two is in production. ▶

Find Out More *(continued)*

◀ *The Elephant Man*, directed by David Lynch John Merrick is affected by a congenital disorder known as Proteus Syndrome. With the help of Dr Frederick Treves he recovers some of his dignity. Based on a true story from the nineteenth century, the film stars Anthony Hopkins and John Hurt. Mel Brooks produced it but kept his name out of the credits to prevent viewers making misleading assumptions about the nature of the film. ■